Towards a new regulatory framework for GM crops in the European Union

Scientific, ethical, social and legal issues and the challenges ahead

Towards a new regulatory framework for GM crops in the European Union

Scientific, ethical, social and legal issues and the challenges ahead

edited by:

Leire Escajedo San-Epifanio

Wageningen Academic
Publishers

EAN: 9789086863020
e-EAN: 9789086868452
ISBN: 978-90-8686-302-0
e-ISBN: 978-90-8686-845-2
DOI: 10.3920/978-90-8686-845-2

Cover photo: Soya bean green leaf closeup,
Salunkwadi, Ambjogai, Maharashtra, India
Image ID: 423358045
Copyright: Karve'

First published, 2017

ESLA SciTech
University of the Basque Country

Fundación BBVA

eman ta zabal zazu

Universidad del País Vasco
Euskal Herriko Unibertsitatea
The University of the Basque Country

This book is a result of a project granted by the *2015 edition of the BBVA Foundation Grants for Researchers and Cultural Creators*. P.R. Leire Escajedo San-Epifanio.

Table of contents

Three decades later, what has become of the EU Regulation of GMOs?

Addressing the scientific, ethical, social and legal challenges ahead from different countries and disciplines

L. Escajedo San-Epifanio
Professor of Constitutional Law and Law & Ethics in the Biosciences, Department of Constitutional Law and History of Political Thought, University of the Basque Country, P.O. Box 644, 48080 Bilbao, Spain; leire.escajedo@ehu.eus

In his book la *Tercera Revolución Verde*, (The Third Green Revolution) García Olmedo stated that 'for more than a hundred years' human beings practised *the art of genetics without knowing it*, adapting and improving those species from which they obtained such essential products for survival as food (García Olmedo, 1999). With the slow passage of time, that domestication of species – in particular of plants – provoked in them *dramatic morphological and physical changes* which would be transmitted to subsequent generations or, in other words, genetic alterations. In this intuitive manner, then, contributions were made to what centuries would become known as Genetics and its applications. Scientific and technological evolution brought about in the 20th century a new focus in the use and manipulation of other living organisms: a biotechnology that incorporated new knowledge and tools into the trajectory of traditional biotechnology. The significance and potential of these new techniques was emphasised, with caution, by numerous national and international bodies. Thus, for instance, chapter 16 of *Agenda 21 'identified the emergence of modern biotechnology as a promising tool through which significant contributions to global sustainable development could be made, from better healthcare and improved food security to enhanced environmental protection and cleaner energy'*, always warning of the need for *'safe and environmentally sound management of biotechnology'*[1]. With different expressions, to refer specifically to those methods that deliberately seek to effect genetic alterations intervening at a molecular level through the use of recombinant gene technology, academic literature has been employing terms such as *genetic engineering* or *genetic manipulation*, although what has prevailed in the sphere of biotechnological politics is a term that, though less precise, in the opinion of many is simpler[2]: *genetic modification*. Hence both in legislation and in public debate the frequent use of the expression *genetically modified organism* (GMO) to refer to living organisms that have been altered in this way.

Aware of the potential associated with modern biotechnology and the need to administer its use in a healthy and sustainable manner, the EU, along with the USA, was one of the first regions in the world to address the question of a legal framework for GMOs. The best-known legal measures of that first legal framework were two Directives of April 23, 1990 on the contained use and deliberate release

[1] UN Department of Economic and Social Affairs. 2012. Draft of *Review of the Implementation of Agenda 21 and the Rio Principles.* Sustainable Development in the 21st Century (SD21).
[2] R. Paarlberg (2003). *Politics of Precaution*. IFPR.

DOI 10.3920/978-90-8686-845-2_0, © Wageningen Academic Publishers 2017

of GMOs respectively (Directives 90/219 and 90/220/ECC[3]). However, the decisions incorporated into the latter and into the legislative texts cannot be understood beyond the context of a participatory process in which from the late 1970s onwards all the sectors involved met and compared, amongst other things, public and private research in the fields of molecular biology, plant biology, ecology and population biology. The objective was that the regulatory framework should provide an adequate channel for bioscientific and biotechnological development, while at the same time clearly reinforcing environmental and health guarantees, as well as the level of protection of consumers' rights to which Member States were committed. Since that date, the European Union has been making significant budgetary contributions to research and development both in agrobiotechnologies as a whole and, specifically, in improving knowledge with regard to the risks associated with these and the best ways of dealing with these risks.

Differing very early on from the one which became established in the USA, the approach employed by the EU in constructing the legal framework for the agri-food use of GMOs had a particular influence upon other countries, especially in the wake of the active participation of Member States and the EU in the United Nations Conference on Environment and Development (UNCED), Rio de Janeiro (1992). This Conference gave rise to – among many other things – a series of important agreements in questions of the environment and biotechnology. Expectations on biotechnology 'must be tempered by realism', reflected the aforementioned chapter 16 of *Agenda 21*, but '*nevertheless, it promises to make a significant contribution in enabling the development of, for example, better health care, enhanced food security through sustainable agricultural practices, improved supplies of potable water, more efficient industrial development processes for transforming raw materials, support for sustainable methods of afforestation and reforestation, and detoxification of hazardous wastes*'. In the light of this potential, *Agenda 21* appeals for the development of principles '*to ensure the environmentally sound development of biotechnology, to engender public trust and confidence, to promote the development of sustainable applications of biotechnology and to establish appropriate enabling mechanisms*'.

Scarcely a decade later, a comparison of the level of biotechnological development in different regions of the world provided evidence that the first legal framework for GMOs in the EU had not succeeded in offering a suitable space for the scientific and technological development of agrobiotechnologies. In spite of the technical quality of the legal instruments implemented, an accumulation of political and economic factors was hindering research and development, especially with regard to the use of GMOs in agriculture and food. As will be shown, according to several trading partners of the EU in the WTO, the rate of commercial cultivation of GMOs in the EU was virtually non-existent and three *types of conduct were hindering imports of agrobiotechnological products:* authorisation applications for new products were not being processed, unreasonable delays were occurring in the finalisation of some applications already in process and some Member States had banned in their territory already approved GM products availing themselves of safeguard clauses.

From that crisis, analysed from different perspectives by several of the chapters of this book, emerged in 2001 what is today the most exhaustive and demanding corpus of legislation in the world in

[3] Directive 90/219/EEC of 23 April 1990 on the contained use of genetically modified micro-organisms and Directive 90/220/EEC on the deliberate release into the environment of genetically modified organisms.

relation to the use of GMOs in agriculture and human and animal foods. Revolving around Directive 2001/18/EC[4] and 2003 EC Regulations 1829 and 1830 on GM food and feed, and its labelling and traceability[5], this legal framework cannot be interpreted independently of Regulation 178/2002, which basically establishes the criteria of quality, safety and consumer protection for the entire EU agri-food market. This Regulation, known as the *EU General Food Law,* established amongst other things the prominent role corresponding to the European Food Safety Authority (EFSA) regarding safety and consumer protection vis-à-vis the agri-food market. This European authority was entrusted with providing independent scientific advice and notifying of any risks associated with the food chain. In fulfilling its role, the EFSA collaborates closely with the European Medicines Agency (EMA), European Chemicals Agency (ECHA), European Center for Disease Prevention and Control (ECDC) and European Environment Agency (EEA).

The general objective of Directive 2001/18/EC was and is to harmonise national legislation on GMOs in the twenty-eight EU Member States in compliance with the precautionary principle and to ensure that individuals and companies take necessary measures to safeguard the environment and human health prior to intentionally releasing into the environment a GMO or placing in the EU market a GMO or GM products. In practice, this translates into the implementation of a regulated and objective procedure that must be passed by any authorisation application for uses involving deliberate release of GMOs into the environment, and specifies different obligations and measures considering two factors: (1) on the one hand, the *nature* of the modified organisms in question; (2) and on the other, the characteristics and objectives of the particular *operation* to be performed. Unlike the Directives of the 1990s, which assigned risk assessment to competent national authorities, Directive 2001/18/EC identifies as a key element in decision-making a scientific evaluation led by the EFSA, an independent and unique scientific authority for the whole EU but in a process in which there are channels for the participation of both Member States and other actors. On completion of the scientific assessment it corresponds to Member States, assisted by the Commission, to adopt a final decision for the EU as a whole. For various reasons, however, this decision-making model has also proven to unsatisfactory, despite having been designed with the aim of consolidating governance of biotechnologies and promoting '*dialogue, coherence and cooperation as regards the implementation of current and future action*' in the words of the Communication of the Commission *Life sciences and biotechnology – A strategy for Europe (2002).*

We can speak of a third legal framework, or at least the process of its construction, from 2010 onwards, especially in the wake of the tension produced between different European institutions with regard to the option of a final decision on *Pioneer's corn 1507* In its Communication *Reviewing the decision-making process on genetically modified organisms* (2015)[6] the European Commission acknowledged that decision-making in regard to GMOs had become '*a notable exception in relation to the normal functioning of the Union*' and that it was essential to adopt a decision in this respect. In

[4] Directive 2001/18/EC on the deliberate release into the environment of genetically modified organisms.

[5] Regulation EC No 1829/2003 on genetically modified food and feed and Regulation EC No 1830/2003 concerning the traceability and labelling of GMOs.

[6] Communication From the Commission to the European Parliament, the Council, the European Economic and Social Committee and The Committee of the Regions *Reviewing the decision-making process on genetically modified organisms* (GMOs), COM (2015) 176 final.

order adequately to evaluate the significance of this affirmation in the context of European Community Law it is worthwhile recalling that the governmental engineering of (today's) EU has encountered, since the origins of the EEC, many tensions and difficulties in its sphere of competences, and will doubtless continue to do so. Consider, for example, the difficulties experienced in only the past two years, in areas such as the control of financial institutions, the crises of Greece or Brexit, and the drama of those displaced by wars. For this reason, and in comparative terms, it is surprising that with the trajectory and the experience of the EU it has not been possible to forge an efficient common management model in relation to a group of products that – in contrast with the over 70 million hectares cultivated in the USA – has in the EU a presence of less than 100,000 hectares throughout the 28 countries and which furthermore has not provoked a single food or environmental alert within the territory.

In order to address this situation the Commission proposed, to put it briefly, an exceptional decision: partially devolve to Member States the capacity freely to adopt, as they saw fit, some of the specific decisions that are however very similar to those which on the contrary are adopted on a daily and harmonised basis at EU level with regard to other products, both agri-food and non-agri-food. In relation to GM crops, this partial devolution of competence was performed via Directive 412/2015 *amending Directive 2001/18/EC as regards the possibility for the Member States to restrict or prohibit the cultivation of genetically modified organisms (GMOs) in their territory*. Implementation of the Directive meant accepting *de ius* (within a legal framework) something that was already happening *de facto*, because many of the 19 Member States that have formally exercised their opt-out right were already restricting *de facto* the use of these crops in their territory.

Within a single market with consolidated instruments of harmonisation and guarantee of food safety, consumer protection and fair competition, legally formalising this diversity is something far more complex than might first appear. In fact, the devolution of competence is only partial, because totally nationalising decision-making GM crops, foods and feeds would have required an *ad hoc* modification of the founding Treaties of the EU. Without fracturing the general framework of governance, on the one hand there is maintenance at an EU level of decisions on risk assessment vis-à-vis GMOs and general authorisation with regard to crops, foods and feeds and, on the other, Member States are granted the possibility of exercising an opt-out option regarding the specific use of GM seeds in their territory. Exercise of that opt-out, warns the Commission, cannot affect the *EU acquis* or the EU's trade commitments with third countries. Among other things, this involves the need for the exercise of this option to be based on *other criteria than those assessed by the EFSA at EU level*. Because this institution fulfils, in relation to the EU's food chain as a whole, a vital function in independent scientific assessment of food quality and safety.

Following the materialisation in practice of this partial devolution of agricultural competences with regard to GMOs, *the next step* in the construction of the new legal framework appeared to be that of according Member States a certain margin to, should they so desire, ban or restrict in their territories trade in GM agri-food products. However, and as we shall see in more detail, the first draft of this proposal for modification of Regulations 1829[7] and 1830/2003[8] was categorically rejected by the EU Parliament (October 2015), among other reasons on account of its incompatibility with the principles of the single market. At the date of publication of this book we are still waiting to see how this proposal will evolve.

Coinciding with the approval of Directive 412/2015[9], in 2015 there began in the University of the Basque Country (Bilbao, Spain) a research project entitled *'A la tercera, ¿va la vencida? Hacia el Tercer Marco Regulador de los OMGs en la UE: Revisión crítica del camino recorrido y retos pendientes'. ('Third time lucky? Towards the Third Legal Framework for GMOs in the EU: a critical review of the road travelled and the remaining challenges').* Participants included people from different disciplines and countries, with diverse and complementary visions and opinions with regard to agrobiotechnologies in general and GMOs in particular. This Project received financial support from the *Fundación BBVA* within its programme of *Grants for Researchers and Cultural Creators.* There were three main issues to be addressed. Firstly, what is the general view with regard to the two generations of legal frameworks which, despite their high technical standard, have not proven satisfactory? Secondly, what has been the legal and political path chosen by those regions of the world where agrobiotechnological crops have become well-established and how are our choices and our characteristics different from those of these regions? Finally, what are the keys in order for this legal framework to be – at the third time of asking – the definitive one that manages satisfactorily to reconcile the different interests expressed in this conflict?

After a period of work in different formats (presentations, reports, workshops, consultations) and in the hope of contributing to the construction of some common ground for the plurality of stances expressed in the debate on the present, past and future of GMOs in the EU, this monographic work was compiled. In it, separate chapters offer the reflections of 14 experts from different disciplines and countries, with very diverse opinions in relation to the questions addressed.

Professor López Basaguren, an expert jurist in the field of Multilevel Governance provides a comparative analysis of the past, present and future of the decision-making models on GM crops in the EU. He explains the way in which gradually and since the late 1980s the European Commission has sought to bring positions closer together with regard to GMOs and how, surprisingly, these efforts seem to have led the EU in the opposite direction: towards fragmentation and disagreement. This legal institutional vision of the regulatory framework for GMOs in the EU is complemented by the contribution of Professor Gonzalez Vaqué, expert jurist in Food Law who served as an advisor to both the FAO and the European Commission. González Vaqué explains the significance of Directive (EU) 2015/412 in the EU Acquis, its scope and significance. Complementing this vision, researcher Romero Caro,

[7] European Union (2003). Regulation (EC) No 1829/2003, genetically modified food and feed, OJ L 268, 18/10/2003. Available at: http://tinyurl.com/y7b255p8.

[8] European Union (2003). Regulation (EC) No 1830/2003 concerning the traceability and labelling of genetically modified organisms and the traceability of food and feed products, OJ L 268. 18/10/03. Available at: http://tinyurl.com/y94hrdwp.

[9] European Union (2015). Directive (EU) 2015/412 of the European Parliament and of the Council of 11 March 2015 amending Directive 2001/18/EC as regards the possibility for the Member States to restrict or prohibit the cultivation of genetically modified organisms (GMOs) in their territory Text with EEA relevance. OJ L 68. Available at: http://tinyurl.com/y7fa2kfq.

also a jurist, offers a study of the practical implementation of the Directive and gives an account of the way in which 19 Member States have made use of the opt-out rights regarding GMO crops: which formulae they have employed for the opt-out, with what content and territorial extension, and upon the basis of what arguments. This block concludes with the contribution of Professor Alicia Arroyo, expert in Commercial Law and Consumer Protection, who explains the legal treatment and labelling requirements of transgenic foods as compared to those of other foods.

Via three different approaches, three works then explore the current situation of GMOs in the EU and the main threads of the debate on the subject. Professor Mertxe de Renobales, Professor of Biochemistry and Molecular Biology and recipient of the International Society of Bioethics 2009 Award, contributes with a *scientific view of the current status of transgenic crops, foods and feed in the EU*. Two subsequent works deal with the ethical issues surrounding GMOs in Europe. Firstly, Matthias Kaiser, director of the Centre for the Study of the Sciences and the Humanities of the University of Bergen (Norway) presents the text *Franken-food or techno-fix? – Ethical issues of GMOs in Europe* and Dr Donald Bruce of Edinethics (United Kingdom) addresses the new contributions made by Genome Editing to ethical-social reflection.

The next three works consider, via different backgrounds and approaches, the regulatory conflicts that have arisen in the EU in relation to GMOs. Particularly relevant is the reflection upon the role in the EU regulatory framework for GMOs corresponding to the EFSA, in its scientific, political-legal and ethical-social dimensions. Doctors Casacuberta, Nogué and Du Jardin address the *Scientific consensus and EFSA assessment of GMOs*, Professor Salvador Martínez – expert jurist on independent authorities – explains the role of the EFSA, namely its functions and competences in relation to other independent authorities, and Professor Levidow, of the Open University (UK) considers from an ethical-social perspective regulatory conflicts regarding GMOs in the EU.

Finally, and in the light of all the contributions to the project, Professor Escajedo San Epifanio, Doctor in law, Doctor in Biological Sciences, and Professor of Constitutional Law, presents as principal researcher of the project a series of conclusions and reflections which, via the history of biotechnological policies in the EU so far, contemplate the future of the legal framework that began with Directive 2015/412. The monograph closes, by way of a comparison from another region, with the work of Professor Pakky Redi, Executive Director of Agrobiotech Foundation India, addressing the ethical and social aspects of agrobiotechnologies in India.

By way of conclusion, we would like to express our gratitude to the Fundación BBVA for believing in this project and supporting it with its programme of *Grants for Researchers and Cultural Creators of 2015*. The Project has been conducted from a perspective of governance and conformation of regulatory frameworks, in other words, via an approach with a significant presence of the constitutional law discipline. But this is far from being a classical theoretical-legal work from the field of Constitutional Law. Jurists and political scientists have participated alongside recognised experts in ethics, agronomy, molecular biology and economic sciences, and different stakeholders have been consulted. It is our hope that with our reflections we can contribute to articulating a legal framework that will facilitate, on the one hand, satisfactory, healthy and sustainable advances in agrobiotechnologies, and, at the same time, a better mutual understanding between the diversity of socio-political and ethical visions that, in this question as in many others, characterise our plural Europe.

1. Decision-making on genetically modified crops in the EU: finding the way out or going even deeper into the maze?

A. López-Basaguren
Professor of Constitutional, Department of Constitutional Law and History of Political Thought, University of the Basque Country, P.O. Box 644, 48080 Bilbao, Spain; alberto.lopez@ehu.eus

Abstract

Within the EU, the decision-making process on GMOs has raised major problems with regard to the requirements of the single internal market. Market unity in the EU entails freedom of movement within the Union; this means that any product that has accessed the market in any of its constituent States (either because it has been produced or prepared there or because it has been imported by a firm established there) may be freely marketed in any of the other states and no state is entitled to restrict that freedom in its territory. However, in the case of GMOs, some states have sought to limit their circulation. This has created a conflict between the restrictive actions of these states and the principles of the internal European market. This chapter analyses the EU's attempts to regulate and harmonise the decision-making process on authorisation of GMOs and products that contain them, as it seeks to uphold the principles of the Single Market while at the same time guaranteeing protection of human and animal health and the environment. It starts by looking at the first two regulatory frameworks for GMOs at EU level. The key regulatory texts key in this regard were Directives 90/219 and 90/220 and 2001/18/EC, both of which failed to achieve their objectives and the chapter examines the reasons for that failure and its consequences. It goes on to analyse the European Commission's approach to the regulatory framework and the inclusion of an opt-out for Member States in the GMO decision-making process. Directive 2015/412, now passed, allows States to prohibit the cultivation of GMOs in their territory. In a proposed amendment to Regulation EC 1829/2003, the Commission recommended that any Member States that so wish might also exclude the use of GMOs in their territories, within certain limitations and provided their grounds were compatible with EU Treaties. This proposed amendment was rejected, and as a result, the regulatory framework on GMOs in the EU currently stands at something of an impasse. The chapter analyses this situation and the fact that surprisingly (and unexpectedly) it was the European Parliament that pointed out the problems with the proposal: that it fails to ensure the necessary legal certainty and adequate tools for Member States wishing to opt-out, does not define the term 'use' and attacks the bases of the internal market.

Keywords: decision making on GMOs, EU integration and GMOs, regulation 178/2002 and GMOs, opt-out in the cultivation on GMOs, opt-out in the use of GMOs

DOI 10.3920/978-90-8686-845-2_1, © Wageningen Academic Publishers 2017

1.1. The legal framework for GMOs in the EU, from the perspective of the Single Market

1.1.1 Economic integration and Single Market

Regulation of GMOs in the EU is strongly determined by the principles that govern European integration; fundamentally, the demands imposed by the principle of market unity, as it has taken shape throughout a singular historic process.

Market unity has been the basic objective of the process supranational European integration since its origins, when the European Communities were created in the 1950s. Over the last two decades in particular, European integration has broadened its objectives in qualitative fashion, intensifying, on the one hand, economic integration, with the establishment of Economic and Monetary Union, and, on the other, with a clear decision to enter further into the political sphere, especially with the Justice and Home Affairs Policy and the Common Foreign and Security Policy. In this process, far from losing importance, the effective guarantee of market unity has a decisive role, to the extent that one might say that today, more than ever it, *'lies at the heart of the EU'* (Barnard, 2004).

The old EEC Treaty (1957) established that 'the aim of the Community' consisted in '*establishing a Common Market and progressively approximating the economic policies of Member States*' (article 2); a 'common market' that would be characterised by '*the elimination, as between Member States, of customs duties and of quantitative restrictions in regard to the importation and exportation of goods, as well as of all other measures with equivalent effect*' (article 3a). The Treaty contemplated achieving market unity upon the basis of *freedoms of movement*: of goods, people, services and capital (Barnard, 2004). This was the application of the 'principle of *negative* integration', which was based upon the conviction that market unity would be achieved, simply, by removing the obstacles to trade (Barnard, 2004).

Reality showed that this was a difficult objective to attain, that it involved considerable difficulty and that it would require the adoption of positive measures that would remove obstacles that the application of the *negative principle* did not itself eliminate. It was necessary to guarantee not only the capacity of goods – products – to circulate freely within the territorial limits of the market, but also that there should be no regulations that, indirectly, might in practice act as barriers that would hinder the circulation of goods or result in unequal opportunities for producers accessing the market.

The achievement of this objective was neither simple nor quick, necessitating a new – and definitive – push in the 1980s, with the European Commission taking centre stage (doc. COM(85)310 (EC, 1985)). December 31, 1992 was set as the deadline for attaining what initially was called an 'internal market' – the term contained in article 2 of the current Treaty of the European Union (TEU) – and, finally, 'Single Market', as the most widely used term. The change in terminology does not mean a change in meaning. The terms are synonymous and express the principle of market unity that was sought; but that terminological transition in some way also reflects an evolution that in practice has allowed for progress towards a market free of internal barriers of any kind, neither explicit nor tacit; neither in law (*de jure*) nor in fact (*de facto*). One can safely say that from the date established by the

European Commission – which coincides with the entry into force of the Treaty of Maastricht – market unity in the EU[10] was fully achieved and guaranteed.

Single Market[11] does not only mean customs union, in other words, the absence of import/export duties between different Member States. It means, moreover, uniform regulation of production conditions for goods entering the market, guaranteeing that regulations that might directly or indirectly affect the preparation of products would not represent obstacles or inequalities for producers accessing the market.

Furthermore, market unity in the EU has since its origins been associated with the guarantee of free competition within the market, to the extent that rules governing competition are a constitutive element of the Single Market (Goyder and Albors-Llorens, 2009).

This panorama which categorically prevents EU Member States from adopting regulations that, directly or indirectly, affect conditions regarding Access to the market, free movement and freedom of competition applies in the EU with regard to any kind of product circulating within the internal market, including of course agri-food products and, among these, those that are or that contain genetically modified organisms (GMOs). With respect to the latter, however, the validity of the principle of market unity is more formal than real. In fact, since the mid-1990s decision-making processes regarding GMOs at community level have seen the generation of a series of dysfunctions in the normal functioning of community institutions, to the extent that, following various attempts to agree upon a common framework, Directive 412/2015 (EU, 2015) eventually accepted with regard to GMOs certain exceptions to market unity. This work analyses the problems faced in the GMO sphere from the perspective of the institutional structure of the European Union. From the point of view of Primary Law, any intention of one or several Member States to establish restrictive regulations vis-à-vis GMOs, both their cultivation and their commercialisation – free circulation – is highly likely to clash with the constitutive principles of the Single Market, unless said restrictions are accepted in a general manner, in other words, at a supranational level throughout the EU. Neither does acceptance of restrictions at an EU level, in turn, have an absolute margin of discretion, given that, on the one hand, community institutions are bound by Constitutive Treaties and the principles contained therein and, on the other, the EU must fulfil the commitments to free trade agreed upon with third countries.

I shall now analyse the tensions which, concerning GMOs and their regulation have arisen between, on the one hand, the intentions of some Member States or of some of their political parties and, on the other, demands arising from the Single Market and its rules. Although I believe this is apparent at first sight, I should like to make the point that the problems concerning the access of the GM crops

[10] In fact, the European Single Market is larger than the EU; it is identified with the *European Economic Area* (EEA), which includes, as well as the EU Member States, Norway, Iceland and Liechtenstein; in addition, Switzerland – in spite of not being a member of the EEA – also forms a part of the Single Market, via numerous bilateral treaties.

[11] At present, the internal market is regulated in article 26 TEU, which establishes that the 'Union shall adopt measures with the aim of establishing or ensuring the functioning of the internal market', which 'shall comprise an area without internal frontiers in which the free movement of goods, persons, services and capital is ensured'.

to the Single Market have been utterly exceptional. As the European Commission indicated in the document in which it explains the reasons leading to the adoption of Directive 412/2015/EC (EU, 2015), the legal framework for GMOs and its evolution are an example of *excepción comunitaria*, which reveals all the limitations, the conflicts and the internal contradictions of the EU.

1.1.2 Single Market: the Union's competences and instruments for their exercise

In accordance with what is established in the Treaties[12], the EU has shared competence with Member States in relation to the internal market (article 4 TEU). This means that States can continue to legislate in this area, but only if the EU has not regulated the matter in question or has decided not to exercise that competence. It is possible, therefore, to find aspects or areas affecting market unity that are regulated by Member States' legislation, because the EU has not regulated them. But if the EU, at a later date, regulates that same question, the Member State regulation cedes precedence[13].

It should be borne in mind, moreover, that the EU has exclusive competence regarding the establishing of the competition rules necessary for the functioning of the internal market (article 3.1.b TFEU), which severely limits the capacity of States to regulate areas of the internal market – of shared competence – in which the EU has not exercised competence.

Finally, in this area, one must take into account the powerful instrument that the Treaty of the Functioning of the European Union (TFEU) places in the hands of the EU institutions, in the framework of shared competencies, given the extraordinary practical efficiency it has demonstrated in the effective guarantee of market unity: the approximation of laws[14]. The purpose of this instrument is to achieve the objectives established in article 26 TEU; in other words, the internal market. To this end, the EU will adopt measures directed towards 'approximation of the provisions laid down by law, regulation or administrative action in Member States which have as their object the establishment and functioning of the internal market'. The approximation of laws has been, without a doubt, the most important and effective instrument at the EU's disposal in guaranteeing market unity.

Regulation of the instrument of approximation of laws allows for a margin of flexibility, contemplating in the same article 114 TEU (Barnard, 2004) some mechanisms that may exceptionally permit Member States to maintain provisional measures that do not conform to Union requirements in the guarantee of market unity. This is the only law that permits Member States to adopt different regulation from that established for the EU as a whole. It is the provision invoked by Member

[12] Currently, the two basic treaties that regulate the EU are the Treaty of the European Union (TEU) and the Treaty of the Functioning of the European Union (TFEU), have replaced the former TEU, (approved in Maastricht and modified on repeated occasions) and the Treaty of the European Community (and other Treaties).

[13] The exercise of competences by the EU is guided, as is established in article 5 TEU, by the principles of conferral, subsidiarity and proportionality (article 5 TEU). That means that the Union shall act only within the limits of the competences specifically conferred by the Treaties; the content and form of Union action shall not exceed what is necessary to achieve the objectives of the Treaties; and, in areas which do not fall within the exclusive competence of the Union, only if and in so far as the objectives of the action cannot be sufficiently achieved by the member States and can better achieved at Union level.

[14] The Approximation of Laws is regulated in article 114 TFEU, former article 95 TEC.

States that have sought to be free of European legislation in relation to GMOs, so it is of immense importance in the analysis of the theme of this work. It is an exception which, as is easily understood, given its contrary nature to a regulation necessary to guarantee market unity in this area, is subject to a strict procedure and to no less strict limitations. The exception contemplated in article 114 TEU permits two types of justification for the exception to a harmonisation measure adopted by the EU. Firstly, it contemplates the possibility of a Member State maintaining a national regulation contrary to a harmonisation measure when this is justified by one of the 'major needs' referred to in article 36 TEU[15] 'or relating to the protection of the environment or the working environment' (par. 4). It also provides for the possibility that, subsequent to the adoption of a harmonisation measure, un Member State might consider it necessary to introduce national provisions based on new scientific evidence relating to the protection of the environment or the working environment on grounds of a problem specific to that Member State arising after the adoption of the harmonisation measure (par. 5).

In either of the two cases, the Member State must advise the European Commission of its intention, justifying the latter, which must be approved or rejected within six months – extendable to another six, after having verified whether or not they are a means of arbitrary discrimination or a disguised restriction on trade between Member States and whether or not they shall constitute an obstacle to the functioning of the internal market. Silence on the part of the European Commission will be considered as acceptance of the exception.

The Treaty indicates that in cases in which a Member State employs this channel to highlight a public health problem in an area that has been the subject of harmonisation, la Commission must immediately valuate the need to propose appropriate measures. In the sphere of EU Food Law, as we know, after the adoption of Regulation EC 178/2002 (EU, 2002) and the creation of the European Food Safety Authority (EFSA), the level of harmonisation and management of public health alerts is such that there is no margin for these type of arguments to justify the creation of national restrictions. If a public health issue with regard to a product or group of products is considered to be genuine, the means to be adopted will be projected throughout the market as a whole and not, therefore, only at a Member State level.

Among the grounds for exception to harmonisation measures in the regulation of an area of the internal market in relation to GMOs, three stand out: those relating to human health, to environmental protection and to scientific developments. In this sense, one should remember the precautionary principle. Provided for in the TFEU with regard to risks to the environment (article 191), although, in practice, as the Commission acknowledges, this is a principle that has wider

[15] The 'major needs' referred to in article 36 TEU are: public morality, public policy or public security; the protection of health and life of humans, animals or plants; the protection of national treasures possessing artistic, historic or archaeological value; or the protection of industrial and commercial property. But in no case 'shall constitute a means of arbitrary discrimination or a disguised restriction on trade between Member States'.

application and also covers consumer protection policy (doc. COM (2000) (EC, 2000)) (Allan, 2000; Escajedo San-Epifanio, 2003)[16].

Scientific development plays an important role. A decision taken at a certain moment on technical grounds with the scientific knowledge of the moment, be it to approve or to reject authorisation, may be modified on the strength of new knowledge. We shall see that one of the keys that has differentiated decision-making models regarding GMOs between the 1990s (first generation) and the 2000s (second generation or second legal framework) refers to the way in which this possibility is addressed and the procedure provided to revise, when necessary, decisions taken.

In conclusion, it should be stressed, in any case, that exceptions require a decision by the Commission, which guarantees market unity, even in the case of acceptance of exceptions by one or more Member States. But these would be very singular aspects, directly related to a certain peculiarity of that Member State, since, as we have seen, in the event of the Member State requesting the exception highlighting a problem – risk – of a general nature, the Commission would have to act in accordance, prompting amendment of the law.

1.1.3 The legislative procedure in the EU: basic clarifications

The EU's legislative provisions and the procedure for their elaboration differ significantly from what is usual in Member States, where the division of powers between institutions is directly reflected in the procedure for drafting laws. In the EU, in accordance with article 288 TFEU, there are three fundamental types of legislative measures: Regulation – which has a general application and is binding in its entirety and directly applicable in all Member States; Directive – which is binding as to the result to be achieved, upon each Member State to which it is addressed, but shall leave to the national authorities the choice of form and methods[17]; and Decision – which is binding in its entirety but specifies those to whom it is addressed being binding only on them.

The procedure for adopting EU legislative acts is very complex. The evolution of the EU and the gradual widening of the powers of the Parliament – which, initially, was a merely deliberative body – have evolved into a decision-making process that came to be known as 'codecision procedure' and which the TEU currently terms 'ordinary legislative procedure' and regulates in article 294 TFEU[18]. Essentially, what is different about this procedure is the participation in the drafting of the most

[16] In this *Communication* the Commission specifies that the recourse to the precautionary principle presupposes: (1) identification of potentially negative effects resulting from a phenomenon, product or procedure; (2) a scientific evaluation of the risk which because of the insufficiency of the data, their inconclusive or imprecise nature, makes it impossible to determine with sufficient certainty the risk in question (p. 14).

[17] The EU institutions, however, do not respect, in many cases, the characteristics of the Directives established in the TEU, but attribute to them a very specific content, which leaves very little room for intervention by States to decide upon the 'form and methods', as a result of which, for their content, they are very similar to the Regulations.

[18] But it should be remembered that the 'codecision procedure' was introduced in 1992, with the Maastricht Treaty, and from that moment on evolved until its present formulation in the Treaty of Lisbon. Which means that when analysing the law prior to its introduction we find rules issued in accordance with the previous procedure, in which the legislative function corresponded, depending on the case, to the Council or to the Commission.

important laws – Regulations and Directives – of the three most important political institutions in the EU: Commission, Council and Parliament. In other words, the legislative function in the EU corresponds, jointly, to the European Parliament and the Union Council, at the request, as a rule, of the European Commission.

The European Commission drafts the law and refers it to the European Parliament and the Union Council – formed by a Government Representative of each Member State. The Parliament states its position with regard to the Commission's proposal, with the amendments it deems appropriate. This is the first Reading by the Parliament. If the EU Council agrees with the Parliament's position at first reading, the law is passed in these terms. But if the Council, at first reading, disagrees with the stance taken by the Parliament, it will communicate its considered position to the Parliament. The latter has a period of three months – extendable by a month – to: (1) express its approval of the Council's position, in which case the law is passed in the terms established by the Council; (2) reject the position established by the Council at first reading, in which case the law proposed is not considered to be passed; (3) propose amendments to the position expressed by the Council at first reading. In the event of Parliament proposing amendments to the text agreed upon by the Council, within three months the Council can adopt them via a qualified majority[19], in which case the law will be passed in these terms. If the Council decides, however, to reject them, totally or partially, the *Conciliation Committee* will meet, a committee formed by an equal number of members of the Council and of the Parliament and in which the Commission participates. Its task is to produce a joint text – which must be approved by the qualified majority of the Council representatives and the simple majority of those of the Parliament – based upon the texts adopted at second reading by the Parliament and the Council. The Committee has three weeks – extendable by two weeks – in which to try to reach an agreement. In the event of agreement being reached, the text must be approved, at third reading, by the EU Council by qualified majority and by the Parliament by simple majority.

As can easily be deduced, this is a procedure that, in the event of disagreement, may be long and complex and even, on occasions, difficult to conclude. In a question in which positions are so far apart as in the case of the use of GMOs in the agri-food sector, a procedure requiring the quest for multiple consensuses and in institutions expressing very different interests may complicate to a considerable degree the taking of decisions. The Commission, as 'guardian of Treaties' will try to defend fulfilment of the objectives of integration, amongst which it will emphasise the guarantee of market unity. But the Member States (or the political forces governing the latter at the specific moments of decision-making) will also serve other interests, which will see them position themselves in favour of or against GMO cultivation and commercialisation in general or of some of these products in particular. In over three decades now of EU biotechnological policies, the Parliament's stance seems to have evolved in erratic fashion. On occasions it has been closer to the Commission, on others, to some Member States; finally, at moments when the Commission seemed to bend itself to the will of States, Parliament, on occasions, has sought to distance itself from both, indicating that this is not the correct path for the Union.

[19] Although it will require the unanimous support of the Council with regard to the modifications proposed by the Parliament in respect of those in relation to which the Commission has issued a negative opinion.

1.2 The first framework for GMOs in the EU and the failure of its decision-making model

The EU established the first legal framework for GMOs in 1990, with the approval of two Directives: Directive 90/219 (EU, 1990a), on the contained use of genetically modified micro-organisms, and Directive 90/220 (EU, 1990b), on the deliberate release into the environment of genetically modified organisms.

The cultivation and marketing of GMOs are included, along with other uses, in what Directive 90/220 (EU, 1990b) describes as deliberate release of GMOs. It was approved with the aim of establishing not only a uniform regulation on this matter throughout the EU, but also as a measure to guarantee a high level of health and environmental protection in the use of GMOs. There being no joint bodies yet in questions of risk assessment, a system is articulated in which the competent authorities of the Member States participate and a system of transparency is established, according to which each State must notify the other Member States and the Commission of the products it intends to authorise and its reasons for doing so. If these raise no objections, the State can authorise the product's entry onto the market and, consequently, the entire EU market.

According to Directive 90/220 (EU, 1990b), if any State or the Commission raised any reasoned objection to this authorisation proposal, within 60 days it had to reach an agreement with the Member State seeking the authorisation. If on expiry of this period it had not been possible to reach an agreement, it corresponded to the Commission to adopt a final decision (article 13) following the procedure specified in article 21. In this procedure, the Commission received the assistance of the Committee, presided over by the representative of the Commission and formed by a representative of each State, which was responsible for analysing and debating the proposal put forward by the Commission and taking a decision within a period, depending on the circumstances, established by its president.

In the event of the opinion of the Committee of article 21, formed according to the rules of adoption of agreements in the Council, being contrary to the Commission's proposal, or in cases in which the Committee did not issue and opinion, the Commission could refer the question to the Council. The latter, in turn, had to decide by qualified majority within three months. In the event of the Council not adopting any decision, the law established that the Commission should proceed to approve the measure proposed by the Member State that processed the authorisation request.

As can be seen, this is a procedure aimed at guaranteeing market unity and in which, when one or more Member States consider that there is no authorisation, certain priority is given to the authorisation of products containing GMOs. The procedure leaves room for one or more Member States to express their opposition, both in the article 21 Committee and in the Council, but in these cases, unless a majority is attained, the law reinforces the mediation role which corresponds to the Commission. If the latter considers that the product does not pose problems with regard to its entry onto the market, this is likely to be the sense of the final decision, although the decision-making capacity of the Commission is subject to certain caveats.

The most significant, with regard to the authorisation procedure for access to the market for GMOs or products containing them is that the authorisation of a product by a Member State – or via the alternative procedure in the event of objection – affects the whole market, as a result of which the State that raised the objection cannot hinder the circulation of that product within its territory. This is what is specified by article 13.5 of the Directive: the product authorised by a State in accordance with the procedure established in the Directive does not require further notification in order to circulate freely throughout the EU territory. What is applied, therefore, is the principle known as 'one-door-one-key' (Van der Meulen and Yusuf, 2013-2014).

In an area such as the agri-food industry, with such differing stances regarding the use of GMOs, it is understandable that this decision-making model should encounter considerable difficulties in its practical application. The Commission filed countless appeals before the European Court of Justice (ECJ), which in judgement declared non-compliance with this Directive on the part of several Member States. These proceedings reflected not only the failure in the application of the Directive, but also, the confrontation between the Commission, set on its fulfilment, and many of the Member States that *de facto* achieved their objective of the law not being effectively complied with (Escajedo San-Epifanio, 2008).

What is the effect of this kind of ruling? The infringement proceeding before the ECJ applies when the Commission [20] considers that a State has failed to fulfil an obligation it has according to the Treaties s. Prior to presentation of the appeal before the ECJ, the Commission must issue the State with a reasoned opinion regarding the non-compliance, with the State responsible for the supposed non-compliance having the possibility to make the allegations it deems appropriate (article 258 TEU). If the Commission, after this preliminary procedure, maintains its position and lodges the appeal before the ECJ and the latter considers, in turn, that in effect, the non-compliance occurred, it will issue a ruling requiring the necessary measures to be taken to put an end to the non-compliance. This is what is known as a 'declarative' Ruling; in other words, the ECJ 'declares' that in effect, a particular Member State is responsible for the non-compliance denounced by the Commission. This type of ruling has kinds of effects. On the one hand, and in a direct manner, from this moment onwards it is the State that is responsible for taking appropriate action to put an end to the situation of non-compliance; which, often, does not occur. In the event of continued non-compliance, and disobedience of the ECJ ruling, the Commission can file a new appeal, requesting the imposition of a fine (article 260 TEU).

However, the most important effect of Rulings declaring non-compliance of an obligation by a State is indirect. The Directives establish a transposition deadline for the legislation by a Member State of the objectives to be attained therein established. ECJ jurisprudence consolidated almost from its beginnings the interpretation that upon expiry of a transposition deadline of a Directive without the Member State of the EU having transposed it into national Law, the Directive is directly applicable to individuals, who may not only demand its application, even in the absence of applicable

[20] Article 259 TEU provides for, similarly, the possibility of the proceedings for noncompliance of obligations by a State being presented by another Member State, but it is usually the Commission, as guardian of the Treaties' which employs this procedure against States that fail to comply with obligations imposed on them by EU laws, and most particularly, Directives.

national Law, but also claim damages in cases in which a request having been processed under a Directive (for example an authorisation of cultivation or marketing of GMOs under Directive 90/220 (EU, 1990b)), a State had not applied it. Therefore, if a company expected to be able to receive authorisation for cultivation or marketing of a product from a State that does not transpose the directive, the legal system of the EU offers it a means of claiming economic damages caused by the lack of decision regarding its request.

Nevertheless, and in the specific case of Directive 90/220/EEC (EU, 1990b), the inability of the Commission to impose generalised compliance with the Directive was such that it ended up producing a kind of *de facto* moratoria (Escajedo, 2008).

Upon confirmation of failure in practical application of Directive 90/220 (EU, 1990b), there begins a process of reform of the legal framework for GMOs in which, as we shall see, in an attempt to achieve the application of a standard authorisation process throughout the EU the Commission tried to conform to what seemed to be the demands of some Member States. This was an interesting process, because it shows how difficult it is to progress when there is major confrontation between the objectives of the Commission and those of the States. The Commission set to work, preparing different reports, prominent amongst which is the *Report on the Review of Directive 90/220/EEC*, explaining what in its opinion were the reasons for the failure of the Directive (Escajedo San-Epifanio, 2008). From this reflection arose the proposal for amendment of Directive 90/220/EEC (EU, 1990b), in 1998 (doc. COM (1998) 85 (EC, 1998)), which would finally be approved as Directive 2001/18/EC.

1.3 The second legal framework for GMOs and its failure

1.3.1 Directive 2001/18/EC

Directive 90/220 (EU, 1990b) underwent two amendments, both regarding modification of the information required in the notification to be presented to the authority of the Member State with competence to grant the authorisation. Directive 94/15 (EU, 1994) modified the information to be provided by those seeking intentionally to release GMOs into the environment for purposes of research or development or any other purpose other than marketing (Annex II of directive). Meanwhile, Directive 97/35 (EU, 1997a) modified the information to be provided by those seeking to introduce onto the market GMOs or any other product containing them (Annex III of directive). These were modifications that did not alter the procedure established. But, meanwhile, another law was passed which heightened the tension between States: Regulation 258/97 (EU, 1997b), on novel foods and novel food ingredients. Designed to address the control of food products which have not yet had a significant presence on the market, this Regulation would affect, in the absence of another more specific law, foods that are or contain GMOs. The Regulation distinguished between two authorisation processes. One simplified, in the event of consideration that there existed on the Market a 'conventional counterpart' to the novel food which could be identified a 'substantial equivalence', and another more stringent for the case in which the novel food did not have such a counterpart on the market. With regard to GMOs, this opened the debate over whether or not a GM product (for instance, bread made with GM corn) could be the counterpart of a conventional product (Escajedo San-Epifanio, 2003). In a way, the tension returns in the wake of a debate that had been dragging on for almost a decade: from the regulatory point of view, to what extent and in what sense is it relevant that a crop, food or feed is GM? The USA decided in the 1990s that the difference between GM and non-GM was not sufficiently relevant to justify exceptional regulation for the former, whilst the EU had indeed established a specific framework for the regulation of cultivation and the marketing of GM products. Was this an obstacle to judging the status of counterpart between a GMO and a non-GMO? The disagreement also led to a moratorium in the practical implementation of this regulation.

On the basis of the Commission's proposal, as has been said, Directive 2001/18 was approved, which amended Directive 90/220 (EU, 1990b) (Rey García, 2006). This would be the key element of the new legal framework, but, in addition, two specific Regulations were approved on GM foods and feeds, regulating, specifically and independently of 'novel foods', both the authorisation and the traceability and labelling of these products.

The legal framework maintains, in the main, the same authorisation process, in the sense that this is a decision-making process in two phases: a scientific assessment, and political decision-making. But the approval of a general legal framework for food safety (Regulation EC 178/2002 (EU, 2002)) and the creation of the EFSA would bring significant practical modifications to the decision-making process, as we shall see later.

The Commission continues to insist upon its attempt to protect the attainment of the Single Market, in the sphere of GMOs too; but begins to flexibilise its position, particularly in relation to the growing complexity becoming apparent in the field of these products. In any case, both the new Directive and in Regulations 1.829 and 1.830/2003 (EU, 2003a,b) (the latter amending Directive 2001/18) maintain the 'one-door-one-key' principle. Authorisation processes are still initiated, as in the previous legal framework, via issue of a request to any Member State and with the possibility of Member States formulating observations. However, the scientific-technical assessment process would no longer be assigned to the competent authority of the Member State, but was transferred to the EFSA, stating that the EU 'has chosen a high level of health protection', because 'the free movement of safe and wholesome food is an essential aspect of the internal market' (justification of Regulation 178/2002 (EU, 2002)). Authorisations are valid for 10 years (barring establishment, in specific manner, of a shorter validity) although there is an *ad hoc* procedure, with a relevant role assigned to the EFSA; whereby any Member State may request the revision of an authorisation that has already been approved. In any case and at most, the maximum validity of ten years of authorisations means that these will have to be subject to revision, providing the opportunity to take into account scientific knowledge and technical developments that have occurred during that period vis-à-vis the impact of that product upon the environment, human health, etc.

It is important to point out that both the Directive and the entire legal framework created around it continue to maintain the objective, unachieved until then, of guaranteeing free circulation in the internal market of any product that met the conditions required for authorisation of its introduction on the market. Directive 2001/18 introduces a *safeguard clause* (article 23), according to which if a Member State, 'as a result of new or additional information made available since the date of the consent', has specific reasons to consider that a product or a component of an authorised product constitutes a risk to human health or to the environment, it may restrict or ban in its territory, in a provisional manner, the use or sale of a GMO or of a product containing it. Really, this is not a novelty, but rather an *ad hoc* reproduction of the safeguard clause with regard to the approximation of legislation, in general, established in article 114 TFEU (former article 95 TEC). Given that reluctant States will attempt to invoke this clause to justify their non-compliance with the Directive, it was decided to specifically include it in the articles so that claims by Member States based on the safeguard clause might be redirected to the decision-making procedure established within it. Thus, the Member State that applies the safeguard clause must immediately inform the Commission and the other Member States with regard to the actions taken and provide grounds for this decision '*supplying its review of the environmental risk assessment, indicating whether and how the conditions of the consent should be amended or the consent should be terminated, and, where appropriate, the new or additional information on which its decision is based*'. The Commission, assisted by the Committee established in Decision 1999/468 (EU, 1999)[21], will adopt a decision within 60 days – extendable, in certain circumstances. This means that the State that invokes the *safeguard clause* has to justify its reasons and base them on technical reports; and, upon these grounds, it will be within the sphere of the Union institutions where the final decision is taken, which guarantees market unity. If the reasons given referred to food security, the EFSA would have a decisive role in that decision-making. In this way, there is an attempt to limit the use of the safeguard *clause* to those cases in which States provide solid grounds for its use; and technical evidence is required, in order that it might be scientifically evaluated and, in the event of it lacking substance, a decision might be justified rectifying the Member State's negative decision (Escajedo San-Epifanio, 2010; González Vaqué, 2017).

1.3.2 Regulation 178/2002 and the centrality of the European Food Safety Authority

The new Directive 2001/18 coincides in time with the approval of a law known in Community jargon as the *General Food Law:* Regulation 178/2002 (EU, 2002), which establishes the principles and general requirements of food legislation and, among other things, creates the EFSA. This law, although it forma a part of secondary law and therefore does not occupy a position of particular strength within the legal system, is the common and general basis of measures applicable to food and feeds both at Community and national level. Various food safety crises, in particular cases like that of bovine spongiform encephalopathy (mad cow disease) and dioxins evidenced the different ways in which each MS carries out its quality analyses and food safety. None of these crises involved GMOs, but distrust of the general regulation of the agri-food sector and political and social tensions regarding GMOs coincided in this scenario. Inevitably, as agri-food products, GMOs will be affected by a General Food Law that emerge in order to re-establish consumer confidence (Commission Green Paper, The General Principles of Food Law in the European Union (COM (97) 176 end).

Apart from establishing the safety standards applicable to all foods circulating in the EU, Regulation 178/2002/EC (EU, 2002) establishes an assessment system centred upon a specialised technical body, an independent authority of a scientific nature: the EFSA (for a detailed account of its function, see Salvador Martínez, this volume). As far as GMOs are concerned, the challenge faced by this authority is crucial: the guarantee that the authorisation system is in fact uniform throughout the internal market depends upon the credibility of the EFSA's decision. Since 1990, for around 12 years, the

[21] The Committee is composed of the representatives of the Member States and chaired by the representative of the Commission (article 5).

strategy was to try to apply the Directive without a technical-scientific instrument that could impose its authority over States' claims. Following approval of the Regulation, the EFSA would compile all the analyses and technical reports upon which the Commission would base its decisions to authorise or not a GMO or a product containing one.

In this Regulation there are two questions that are particularly worth noting. Firstly, the introduction of the concept of 'risk analysis', which would henceforth prove crucial. Risk analysis would become the basis for the decision to authorise or ban a certain organism or product containing it. Secondly, the way in which the 'precautionary principle' is underlined with regard to environmental effects to determine the authorisation or negation of the introduction on the market of an organism or product of these characteristics. Upon these two elements the EFSA would have to base its actions, in completing its reports, and the Commission, in adopting its decisions to authorise or to deny.

Altogether, the combined application of the provisions included in Directive 2001/18/EC – the contents of which are basically directed towards environmental risk management – and those in Regulation 178/2002 (EU, 2002) and in Regulations 1829 and 1830 on GM foods and feeds, serve to strengthen Commission capacity. This is particularly true in cases in which in accordance with scientific knowledge and applying the precautionary principle, there is thought to be no reason for a Member State to invoke the safeguard *clause* or give reasons that do not reflect the provisions of the clause. Contemplated from the perspective of achieving and guaranteeing market unity this legal framework cannot be questioned. And this is of particular significance in a sector in which the actions of some Member States were repeatedly clashing with the foundations of the internal market.

At this stage it is worth making an important point, which will be revisited with regard to the 2015 Directive. Since its creation, the role attributed to the EFSA with respect to GMOs would not differ from that assigned to it in the General Food Law with regard to the agri-food market as a whole. Hence the striking nature of the contrast between, on the one hand, the considerable recognition of the EFSA's work over the past ten years in the agri-food sector in general and, on the other, the fact that some of the harsh criticism of this institution has been related to GMOs. In order to understand the latter, one must bear in mind that, due to its scientific-technical role, the EFSA has found itself, as it were, in the crossfire. Its rulings have proved decisive when the Commission has, in some cases, prevented simply obstructive actions by States and, in others, extended to the entire internal market a well-founded proposed safeguard presented by a Member State. It is, therefore, a key player in the quest to ensure market unity, guaranteeing, at the same time, in the most efficient way, respect for human health and the environment. In other words, if there is some reason for denying a product access to the market, this should be performed for the internal market as a whole, so that all may benefit from greater and better protection; but if there are no solid reasons for denying a product access to the market, States cannot act in a manner that will break market unity.

It should be borne in mind that there is another very important aspect of European market unity vis-à-vis GMOs: laws regarding imports of products from other countries with which the EU is connected via trade agreements, as well as the EU's observation of WTO regulations. This aspect is particularly important, with regard to imported food and feeds. In accordance with these obligations, there are products containing GMOs import of which to the EU cannot be prevented. Thus, it is possible to find that certain EU Member States have resisted – or prevented – authorisation of the cultivation or commercialisation of a product because it contained GMOs when, however, this is a product that has been authorised in another Member State or imported by any company located in any Member State, even in the one that blocked the authorisation. In both cases, the free circulation of that product throughout the internal market cannot be hindered by the Member State that, however, its cultivation or access to the market in its territory. These situations show the contradiction between the resistant attitudes of certain Member States and the basic principles of the internal market. It is a paradox also revealed in the separation between authorisation of cultivation and authorisation of Access to the market, which has been sought, even by the European Commission itself, as we shall now see. And it poses a significant problem with regard to the Single Market, insofar as it establishes a double standard in relation to GMOs or products containing them: one standard regarding cultivation and another regarding commercialisation.

The authorisation process of Directive 2001/18 permits the authorisation to establish the use or specific uses for which authorisation is granted, excluding others. In other words, it leaves open the possibility of authorising of some GMO crops on EU territory (its industrial use, or the possibility of applying for its use as food or feed, or food ingredient, via Regulations 1829 and 1830/2003 (EU, 2003a,b)), without its use also including the possibility of cultivation. In fact, as of 2015, 69 varieties of GMOs were authorised for different uses in accordance with Directive 2001/18, but among these uses only 6 varieties could be cultivated. This leads us to a major paradox, which is particularly significant: a product may lack authorisation for cultivation in the EU and, however, have no restriction upon its commercialisation in the internal market. This places us on the threshold of the conflict that flared up in late 2015 within the EU between the Parliament and the Commission.

1.3.3 The crisis of the second legal framework

The second legal framework also failed due to the resistance of some Member States to authorise products containing GMOs; especially, to authorise their cultivation, this being the area where the resistance of the State had the potential to be more effective. The European Commission acknowledged this failure in a very important document: *Communication from the Commission on the freedom for Member States to decide on the cultivation of genetically modified crops*[22]. This document would open the process for the reform of Directive 2001/18 and of Regulation (EC) n. 1.829/2003, the two elements that regulated the area of products containing GMOs, one in a general manner and the other in a specific way foods and feeds (Morris and Spillane, 2010; Weimer, 2011). The Commission, on the one hand, communicated its complaints, its displeasure, regarding the behaviour of some States, which was resulting in the failure of the second regulatory framework. It recalled that the EU Council, in December 2008, considered the legal framework for GMOs to be practically complete, to the apparent satisfaction of all. However, the Commission recalled that since adoption of the then-valid legal framework, in a period of six years, seven Member States had banned or restricted the cultivation in their territories of different products containing GMOs invoking safeguard measures. It added that on four separate occasions, the Council had rejected by qualified majority all the Commission's proposals to reject national safeguard measures on GMO

[22] This is a document contemplated, initially, in doc. COM (2010) 375 and which found continuity in doc. COM (2010) 380.

cultivation 'even though in all cases EU scientific assessments had concluded that these measures were not based on new or additional scientific information since the authorisations were granted and therefore such measures were not justified from a legal point of view'. We have seen that this was the mechanism that had been incorporated to overcome the opposition of Member states that was not justified; but this required the Council's support for the Commission's proposal; support which, as the Commission noted, had not materialised. Without agreement between the Council and the Commission the entire building was falling down, enabling reticent Member States to impose in practice their position. There are States that use the safeguard clause employing technical-scientific arguments that EFSA considers to be unfounded. And yet the Council instead of backing the Commission, imposing the Commission's decision, has allowed these Member States to violate the law, breaking the internal market in this field. Nearly half the Member States – 13 out of 27 – invited the Commission to draft a proposal so that Member States might freely decide with regard to GMO cultivation. Accepting a decision-making margin of such characteristics would mean, from the legal point of view, accepting an exception to the inner market in relation to agri-food products that are or that contain GMOs (Klika *et al.,* 2013).

In September 2009, the Commission presided over by Barroso, a very weak Commission, subservient to the interests of the States, agreed to invoke the subsidiarity principle – presupposing that a general regulation was not necessary throughout the EU, but that the Union's objectives could successfully be attained by leaving their regulation in the hands of States. It was admitted that there cannot always be a suitable balance between the EU framework and the need to take into consideration diversity in a Union of twenty-seven States. Whilst given the levels of tension that have been generated vis-à-vis GMO crops and foods this acceptance might be seen as understandable, the fact is that in the Community acquis as a whole it is more than noteworthy, surprising – in the framework of the internal market. Nonetheless, it highlights just how strong can be the opposition of States. Without the participation of the latter, which are represented on the Council and have a very important role in the EU legislative process, it is not possible to achieve the objectives established in the Treaties. The Commission, therefore, in a way accepts its incapacity to impose market unity in this field and the practical ineffectiveness of the *ad hoc* mechanism for administering safeguard clauses that it had introduced in the legal framework for GMOs.

Based on this assumption, there began a process to seek to define a system of GMO authorisation combining, on the one hand, decisions based on scientific data to be adopted at EU level, and on the other, a certain leeway so that Member States could decide with regard to their territory some of the practical consequences of the decision taken at a European level. The idea was to separate, specifically, the decision as to which GMOs were ideal for use in the EU, which would be adopted at Community level for the market as a whole, and certain decisions regarding the specific use that could be made of GMOs authorised within a national territory (in particular, agricultural use). With this approach, the Commission aims to '*to clarify how to implement the freedom for Member States through an approach that combines a revision of the existing recommendation on coexistence, recognising that Member States need more flexibility, with a modification of the existing legislative framework*', accepting '*the possibility for Member States to restrict GMO cultivation from large areas of their territory to avoid the unintended presence of GMOs on conventional and organic crops (GM-free areas)*'. In other words, the Commission will propose an 'opt-out' system so as '*to combine a European Union authorisation system, based on science, with freedom for Member States to decide whether or not they wish to cultivate GM crops on their territory'* (Biszko, 2011-2012). The Commission, in any case, will attempt to impose certain conditions: '*this possibility has to rest on the demonstration by Member States that for those areas other measures are not sufficient to prevent the unintended presence of GMOs in conventional or organic crops*'; and '*the restriction measures need to be proportionate to the objective pursued (i.e. protection of particular needs of conventional or organic farming)'* (Moore Smith, 2011-2012; Poli, 2010).

The crisis of the second legal framework, already assumed with the start of the legal revision process described above, was rendered apparent in even clearer – and shameful for the Commission – fashion with the General Court (former Court of First Instance) Ruling of September 26, 2013, on corn 1507 (Weimer andPisani 2014). In this Ruling the Commission was condemned for *noncompliance with its obligations*; specifically, for not having submitted for consideration by the Council the motion for a resolution on the authorisation of a deliberate release of a GMO requested by Pioneer Hi-Bred in 2001. There was a favourable report on the product by the competent authority of a Member State (Spain) and successive reports in a similar vein from the EFSA. But when the Spanish authority notified the other Member States of its proposal, some expressed their opposition, without the Committee created by Directive 2001/18 obtaining the necessary majority. In accordance with this law, the Commission should refer its proposal to the Council, so the Council could adopt a decision. The Commission, seeing the opposition of several Member States, did not refer any proposal, maintaining a *de facto* moratorium which – according to the USA, Argentina and Canada, in their appeal before the WTO – had existed in the EU since at least 1997.

The ECJ condemned the Commission for noncompliance with its obligations and after the Ruling; the Commission referred the proposal to the Council of the EU (Commission information note, 6.11.2013). However, and in spite of having seven positive scientific assessments from the EFSA, the proposal did not attain a qualified majority in the Council, neither in favour of nor against the Commission's proposal. In accordance with what is established in article 5 of Council Decision 1999/468 (EU, 1999), referred to by article 30(2) of Directive 2001/18, the absence of a qualified majority obliged resolution of the situation via a decision by the Commission (Council press release of February 11, 2014). But this is not what occurred. The pressure placed upon the Commission was considerable, led by various Ministers (i.e. members of the Council, representing their respective Member States) and also exerted by the European Parliament. The latter expressly requested that the Commission disregard the obligation to resolve imposed by Directive 2001/18/EC and compliance with which had been required of it by the ECJ in the ruling referred to (European Parliament. Resolution of 16 January 2014 (EP, 2014))[23]. The Commission – *that* Commission – was incapable of addressing this situation and, in parallel, intensified the preparation work for a new legal framework for GMOs. Directive EC412/2015, the result of this process, consolidated the subdivision of the GMO authorisation process into two stages: differentiating between the general authorisation of a GMO within the EU, as a means to its generic commercial access, and the specific authorisation for cultivation of said GMO in the territory of a Member State. This second decision, that of authorising in a territory the specific cultivation of a GMO previously authorised by the EU was left in the hands of the Member States (Pascalev, 2012; Weimer and Pisani, 2014). At the time of publication of this

[23] This resolution was approved with 385 votes in favour, 201 against and 30 abstentions.

work, corn 1507 has yet to obtain authorisation in the EU and the European has published a second resolution, calling on the Commission to oppose this crop[24].

1.4 The third legal framework or the insistence upon penetrating further into the maze: from the opt-out in cultivation to the opt-out in the use of GMOs

As has already been noted, the Commission published, in the spring of 2015, its proposal for revision of the second legal framework (EC, 2015a), in the light of the failure of the second attempt, in which it sought to accommodate the States' demands that they be granted the capacity to restrict the use of GMOs and the Commission's obligation to the real and effective validity of the principle of unity in the internal market, in accordance with what is established in the Treaties (Pascalev, 2015). The Commission states that 'there have been dysfunctions in the normal functioning of the EU institutions' in relation to the validity of internal market laws in the area of GMOs. What the Commission is acknowledging, in short, is its inability to impose compliance with the two successive legal frameworks in the question of organisation of GMOs.

Under these circumstances, it seeks to respect the minimum terms that the EU Treaties and, specifically, the EU General Food Law establish as indispensable for the correct functioning of the internal market: ensure a high level of health protection and protection of consumers' interests with regard to food; guarantee risk control based on the best scientific knowledge available and with application of the precautionary principle; and avoid obstacles or restrictions to free trade that are incompatible with article 114 TEU and those that might mean noncompliance with international obligations with third countries, including the rules of the WTO.

In an attempt to achieve its objective, the Commission gives way to the Member States in the decision, above all, related to the authorisation – or ban – on the effective cultivation of GMOs in the territory of a Member State. In so doing, the Commission retracts what was characteristic in the initial period of the European Commissions, when market unity was deployed, almost exclusively, in freedom of movement, in other words, in the area of commercialisation, permitting restrictions to the production of GMOs in a Member State. Nevertheless, in the aforementioned framework of the Treaties and international commitments, that restriction on national production does not give the Member State the possibility of preventing in its country the commercialisation of GMOs use of which has been authorised by the EU and that have been produced in another Member State or imported by any company, even one located in the very Member State that prohibits its cultivation.

We shall shortly see the way in which the law finally passed upon the basis of the Commission's proposal includes the latter's pretensions. But, previously, it is necessary to refer to a formal question that, however, has considerable political significance, in institutional relations within the EU, and important legal consequences. The Commission, in document COM (2010) 380 (EC, 2010), in which it accedes to the demands of States accepting that they could prohibit in their territory the

[24] European Parliament Resolution of 6 October 2016 on the draft Commission implementing decision concerning the placing on the market for cultivation of genetically modified maize 1507 (DAS-Ø15Ø7-1) seeds (D046172/00 – 2016/2920(RSP) (EP, 2016)).

cultivation of GMO products authorised for their commercialisation in the EU[25], indicated that the legal revision could '*be done through the inclusion of a new Article 26b in Directive 2001/18/EC and would be applicable to all GMOs which have been authorised for cultivation in the EU, being under Directive 2001/18/EC or under Regulation (EC) no. 1829/2003*' (Poli, 2013).

Upon this basis, the new Directive 2015/412 (EU, 2015) was approved, granting Member States the possibility to restrict or prohibit GMO cultivation in their territory. As the Commission had proposed, this Directive modifies article 26 of Directive 2001/18, enabling States to prohibit GMO cultivation on the following grounds: (1) environmental policy objectives; (2) town and country planning; (3) land use; (4) socioeconomic impacts; (5) avoidance of GMO presence in other products without prejudice to Article 26a; (6) agricultural policy objectives; (7) public policy' (In another chapter of this volume, Romero Caro looks in detail at which Member States have made use of the opt-out and how). It also establishes a period for reviewing the need to adopt measures to avoid cross-border contamination between countries that cultivate GMOs and those that have exercised their opt-out option.

As a final, very recent element in the evolution of the third legal framework for GMOs in the EU, the Commission has presented new Communication regarding the revision of the decision-making process on the 'use' – access to the market – of GMOs (EC, 2015a), Specifically, and in line with the terms that led to the approval of Directive 412/2015, the Commission has proposed the modification of Regulation 1829/2003 regarding transgenic food and feeds (EC, 2015b), the other leg of the legal framework. In this Communication the Commission again describes the situation since 2010, although it adds a series of relevant data. It emphasises that opposition to GMO cultivation has grown amongst both citizens and Member States, and notes that the amount of cultivation is very small, though it underlines the considerable difference between the consumption of GMOs in human food, where it is generally rejected, and the significant use made of GM feed in animal foodstuffs.

The Commission's proposal is '*to amend the 2003 Regulation in such a way as to allow Member States to restrict or prohibit the use, on part or all of their territory, of GM food and feed authorised at EU level for compelling reasons other than the risk to human or animal health or to the environment – that is, criteria other than those assessed by EFSA in its risk assessment*'. In other words, it proposes that the opt-out approved in relation to crops in Directive 412/2015 (EU, 2015) is extended to trade in GM food sand feed. This right to opt-out of trade, in compliance with the provisions of art. 114 TEU, would also make it necessary '*to provide justification for that specific case, taking into account the GMO in question, the type of measure envisaged, and the specific circumstances present at national or regional level that constitute the grounds for such an opt-out*'. The Commission also recalls that '*when exercising this new competence, Member States remain fully bound by their international obligations, including WTO rules*'. The Commission indicates that it aims to safeguard the standard authorisation procedure with regard to the grounds for protecting human and animal health and the environment, guarantee of which corresponds to the EFSA; but in its proposal the Commission only introduces

25 The Commission declared that '*it appears appropriate to amend EU legislation in order to provide in the EU legislative framework on GMOs an explicit legal base to authorise Member States to restrict or prohibit the cultivation of all or particular authorised GMO in part or all of their territories on the basis of their specific conditions*' (p. 6).

two criteria in the decision of States to restrict or ban access to the market of GMOs: that the decision be '*(1) reasoned and based on compelling grounds in accordance with Union law which shall, in no case, conflict with the risk assessment carried out pursuant this Regulation; and (2) proportional and non-discriminatory.*'

This is a really surprising proposal, after all that has been said about the principle of market unity, and difficult to implement in practice in a region in which internal trade barriers have been rendered more flexible so as to guarantee that unity. Far from resolving the conflict between Member States and the difficulty in administering the grounds alleged to restrict or ban GMOs in their territory, this form of acting transfers an eminently political decision to a later moment, when, very probably, the ECJ will ultimately have to assume a leading role. In short, and provided the corresponding demands are presented, it will correspond to the ECJ to analyse whether the State that restricts or bans GMOs does so in accordance with the principles established in the Treaties. The Commission appears to be eluding the responsibility that in EU Food Law as a whole and, specifically, in the two previous legal frameworks for GMOs, corresponded to it. When it acts in this manner, it is difficult to anticipate, in the event of this proposal prospering, whether, in the case of a Member State restricting or banning a GMO that has entered the internal market, the Commission will turn a blind eye and allow States to act as they see fit, or will appeal against the restrictive or prohibitory measures before the ECJ. In the first case, the internal market in this sector will have fallen apart; in the second, it will assign to the ECJ a responsibility, a burden vis-a-vis Member States which the Commission has been unable to assume.

At the time of writing, this proposal has not prospered. Conspicuously, it has come up against an obstacle of considerable significance: the outright opposition of the European Parliament. Important, on the one hand, for the critical role played by the Parliament in passing any legal measures but also, in addition, for the wisdom of the objections made in its resolution. As we have seen, Parliament had previously maintained a reticent attitude with regard to GMOs, especially evident in its stance regarding the conflict generated over corn 1507 and expressed its support of recognition of Member States' opt-out capacity with regard to GMO cultivation, backing the proposal of Directive 2015/412 (EU, 2015).

The European Parliament rejected, at first reading, the proposal for modification of Regulation 1829/2003 (EU, 2003a), asking the Commission to present a new proposal (Legislative Resolution of 28 October 2015). The *Report* accompanying the Parliament's draft Resolution of rejection (Rapporteur: Giovanni La Via) expresses all too clearly the problems presented by the proposal for modification of the Regulation presented by the Commission. The proposal certainly provokes '*serious concerns about the far-reaching consequences of the proposal for the functioning of the Internal Market for food and feed and for the competitiveness of the Union's agricultural sector*'. In relation to this, '*the proposal is likely to have negative indirect effect on imports*', as a result of which it '*may seriously endanger livestock production and also negatively affect agriculture in the EU*'. Furthermore, '*the proposal [is considered] to be almost impossible to implement as border controls no longer exist within the EU's agricultural sector*'; the reintroduction of such controls between Member States '*would reverse the economic achievements of the European Customs Union and the Single Market*'.

Its application, meanwhile, poses serious practical problems. As the Parliament indicates, '*it is unclear how Member States making use of an 'opt-out' will, in practice, ensure compliance with such measures on their territory*'. Finally, the highlights one of the great problems with the proposal. The Commission attempts to maintain the decisive role of the EFSA in the control of grounds related to the protection of human and animal health and of the environment, insofar as safety must be guaranteed throughout the Union: if there are grounds for questioning safety, they should be applied throughout the Union; but if there are not, they cannot be invoked as a pretext for other motives. For this reason, the Commission wants States that avail themselves of the opt-out regarding use justify this decision citing different grounds to those that must be guaranteed throughout the Union. But what are these *other possible grounds*, which must be based on the Treaties and respect the internal obligations of the EU? As Parliament correctly states, 'by stating that Member States must provide 'compelling reasons' to justify their opt-out while respecting the Internal Market and WTO trade rules, the proposal fails to ensure the necessary legal certainty and adequate tools for Member States that wish to lawfully restrict or prohibit the use of GM food and feed on their territory'. All the more so when, as Parliament indicates, '*the Commission has, to the detriment of legal certainty, not provided any definition of the term 'use' in its proposal*'.

It comes as a surprise that Parliament, much closer to purely political considerations related to social sensibility, should now be the EU institution that rubs salt in the wound and warns of the implications that a commercial opt out commercial with regard to the use of GMOs might have for the Single Market and of the consequences that it might have for legal certainty in the very dark terrain in which the Commission appears to situate justification of the commercial opt-out. This is an attitude more befitting the Commission, to which it assigns – as we have seen – the role of guardian of the Treaties, than Parliament, which responds to directly political reasons.

1.5 Conclusions

In such a context, it is hard to know how the issue of the use of GMOs or products that contain them will end in the EU. In any case, it is difficult to justify the current situation, in which there exists a law which is systematically infringed and, at the same time, a *de facto* moratorium, which poses serious problems with regard to WTO laws and the EU's internal obligations. The conflicting interest regarding GMOs prevent coherent government in the EU sphere and reveal, as we have seen, significant risks of fracture of the internal market. Definitively to exclude these products from the internal market, would lead to questioning of administration of borders and the search for some way for each Member State to respond, in this specific case, to the international commercial obligations that might be fulfilled by its actions. Perhaps for this reason, the path taken has been to admit exceptions that, initially appear less drastic, but in the medium and long term will oblige the EU to adopt a definitive decision. In the current situation, there are products use of which has been permitted throughout the market, but which cannot be cultivated in any Member state, and as a mechanism formally to guarantee compliance with article 114 TEU Member States are symbolically asked to 'reason their opt-out'. For the time being, and formally, every opt-out has been admitted, even those that have been accompanied by hardly any reasons at all, but it remains to be seen whether any appeal is lodged against them and what will be, where appropriate, the ECJ's stance vis-à-vis this situation. The moment will come when the EU has to face its other trading partners and deal with the consequences of this legal framework, which may be the source of major conflict between the WTO and the EU.

For the moment, though it seems that the Commission's goal was to try and find a way out of the maze of decision-making processes in the EU with regard to GMOs, the decisions taken by the Community institutions in this area are heading in the opposite direction, leading deeper and deeper into the maze and moving away from a way out in accordance with the Treaties and the principles of the Union.

Acknowledgements

Research Group IT743-13 of the Basque Government's University System and Multilevel Constitutionalism Group (University of the Basque Country, Bilbao -Spain). This paper is a part of the Research Project DER 2013-47774-P, funded by the Spanish Ministry of Economy and Competitiveness.

References

Allan, K.B. (2000). A commentary on the European Commission's communication on the precautionary principle. The Law Teacher 34: 215-221.

Barnard, C. (2004). The substantive law of the EU. The four freedoms, Oxford University Press, Oxford, UK.

Biszko, K.M. (2011-2012). A house divided: exploring implications of decentralized regulation of genetically modified crops in the European Union. Georgia Journal of International and Comparative Law: 527-554.

Council of the European Union (2008). Council of the European Union-environment of 4 December 2008 on Genetically Modified Organisms – GMOs –, AGRILEG 223; ENV 961. Brussels, Belgium.

Council of the European Union (2014). Press release of the EU Council – general affairs, 11 february 2014. Brussels, Belgium. Available at: http://tinyurl.com/jx7rhop.

Court of Justice of the European Union (2013). General Court (seventh chamber), judgment of 26 September 2013 – Case T-164/10 – Pioneer Hi-Bred International. Brussels, Belgium.

Escajedo San-Epifanio, L. (2003). Revocación de autorizaciones relativas a OMG al amparo del principio de precaución (con especial atención a la doctrina del TJCE). Revista de Derecho y Genoma Humano – Law and the Human Genome Review 18: 139-161.

Escajedo San-Epifanio, L. (2008). Biotechnologie, santé et environnement dans l'Union européenne. Aspects politiques et juridiques. Revue du Marché Commun et de l'Union Européenne 517: 255-266.

Escajedo San-Epifanio, L. (2010). Vers un cadre régulateur de la sécurité de l'application de la biotechnologie: le chemin parcouru. Revue Générale du Droit International Public 1: 5-34.

European Commission (EC) (1985). White paper completing the internal market, document COM(85)310. EC, Brussels, Belgium.

European Commission (EC) (1998). Proposal for a European Parliament and Council Directive amending Directive 90/220/EEC on the deliberate release into the environment of genetically modified organisms, COM(1998) 85 final. Official Journal of the European Union C 98: 139.

European Commission (EC) (2000). Communication from the Commission on the precautionary principle, document COM (2000)1 final. EC, Brussels, Belgium.

European Commission (EC) (2010). Document COM(2010) 380 final, of 13.07.2010: Communication from the Commission to the European Parliament, the Council, the Economic and Social Committee and the Committee of the Regions on the freedom for Member States to decide on the cultivation of genetically modified crops. EC, Brussels, Belgium.

European Commission (EC) (2015a). Communication from the Commission to the European Parliament, the Council, the European Economic and Social Committee and the Committee of the Regions. Reviewing the decision-making process on genetically modified organisms (GMOs), doc. COM (2015) 176 final. EC, Brussels, Belgium.

European Commission (EC) (2015b). Proposal for a Regulation of the European Parliament and the Council amending Regulation (EC) no. 1829/2003 as regards the possibility for the Member States to restrict or prohibit the use of genetically modified food and feed on their territory, doc. COM(2015) 177 final (2015/93 COD). EC, Brussels, Belgium.

European Parliament (EP) (2014). Resolution of 16 January 2014 on the proposal for a Council decision concerning the placing on the market for cultivation, in accordance with Directive 2001/18/EC of the European Parliament and of the Council, of a maize product (Zea mays L., line 1507) genetically modified for resistance to certain lepidopteran pests (2013/2974(RSP). EC, Brussels, Belgium.

European Parliament (EP) (2016). European Parliament Resolution of 6 October 2016 on the draft Commission implementing decision concerning the placing on the market for cultivation of genetically modified maize 1507 (DAS-Ø15Ø7-1) seeds (D046172/00 – 2016/2920(RSP)). EP, Brussels, Belgium.

European Union (EU) (1990a). Council Directive 90/219 of 23 April 1990, on the contained use of genetically modified micro-organisms (90/219/EEC). EU, Brussels, Belgium.

European Union (EU) (1990b). Council Directive 90/220 of 23 April 1990 on the deliberate release into the environment of genetically modified organisms (90/220/EEC). EU, Brussels, Belgium.

European Union (EU) (1994). Commission Directive 94/15/EC of 15 April 1994 adapting to technical progress for the first time Council Directive 90/220/EEC on the deliberate release into the environment of genetically modified organisms. Official Journal of the European Union L 103.

European Union (EU) (1997a). Commission Directive 97/35/EC of 18 June 1997 adapting to technical progress for the second time Council Directive 90/220/EEC on the deliberate release into the environment of genetically modified organisms. Official Journal of the European Union L 169.

European Union (EU) (1997b). Regulation (EC) no. 258/97 of the European Parliament and the Council of 27 January 1997 concerning novel foods and novel food ingredients. Offcicial Journal of the European Union L 43.

European Union (EU) (1999). Council Decision 1999/468/EC of 28 June 1999 laying down the procedures for the exercise of implementing powers conferred on the Commission. Official Journal of the European Union L 184.

European Union. (EU) (2002). Regulation (EC) no. 178/2002 of the European Parliament and the Council of 28 January 2002 laying down the general principles and requirements of food law, establishing the European Food Safety Authority and laying down procedures in matters of food safety. Official Journal of the European Union L 31.

European Union (EU) (2003a). Regulation (EC) no. 1.829/2003 of the European Parliament and of the Council of 22 September 2003 on genetically modified food and feed. Official Journal of the European Union L 268.

European Union (EU) (2003b). Regulation (EC) no. 1.830/2003 of the European Parliament and of the Council of 22 September 2003 concerning the traceability and labelling of genetically modified organisms and the traceability of food and feed products produced from genetically modified organisms and amending Directive 2001/18/EC. Official Journal of the European Union L 268.

European Union (EU) (2015). Directive (EU) 2015/412 of the European Parliament and the Council of 11 March 2015 amending Directive 2001/18/EC as regards the possibility for the Member States to restrict or prohibit the cultivation of genetically modified organisms (GMOs) in their territory. Official Journal of the European Union L 68.

González Vaqué, L. (2017). Directive (EU) 2015/412 in the EU Acquis: its scope and significance. In: Towards a new regulatory framework for GM crops in the European Union, edited by L. Escajedo San-Epifanio. Wageningen Academic Publishers, Wageningen, the Netherlands.

Goyder, J. and Albors-Llorens, A. (2009). Goyder's EU competition law, 5th edition. Oxford University Press, Oxford, UK.

Klika, C., Kim, J. and Versluis, E. (2013). Why science cannot tame politics: the EU comitology rules and the centralised authorisation procedure of GMOs. European Journal of Risk Regulation 4: 327-334.

Moore Smith, L. (2011-2012). Divided we fall: the shortcomings of the European Union's proposal for independent Member States to regulate the cultivation of Genetically Modified Organisms. University of Pennsylvania Journal of International Law 33(3): 841-870.

Morris, S.H. and Spillane, C. (2010). EU GM crop regulation: a road to resolution or a regulatory roundabout? European Journal of Risk Regulation 1: 359-369.

Pascalev, V. (2012). Can science tame politics: the collapse of the new GMO regime in the EU. European Journal of Risk Regulation 3: 190-201.

Pascalev, V. (2015). GMO regulation in Europe: undue delegation, abdication or design flaw? European Journal of Risk Regulation 6: 573-578.

Poli, S. (2010). The commission's new approach to the cultivation of Genetically Modified Organisms. European Journal of Risk Regulation 1: 339-344.

Poli, S. (2013). The member states' long and winding road to partial regulatory autonomy in cultivating genetically modified crops in the EU. European Journal of Risk Regulation 4: 143-157.

Rey García, P. (2006). Directive 2001/18/EC on the Deliberate Release into the Environment of GMOs: an overview and the main provisions for placing on the market. Journal for European Environmental and Planning Law 3: 3-12.

Van der Meulen, B. and Yusuf, N. (2013-2014). One-door-one-key principle: observations regarding integration of GM authorization procedures in the EU. Penn State Law Review 118: 877-892.

Weimer, M. (2011). What price flexibility? The recent Commission proposal to allow for national 'Opt-Outs' on GMO cultivation under the deliberate release directive and the comitology reform post-Lisbon. European Food and Feed Law Review 5: 232-239.

Weimer, M. and Pisani, G. (2014). The EU adventures of 'Herculex'. Report on the EU authorization of the Genetically Modified maize 1507. European Journal of Risk Regulation 5: 208-212.

2. Directive (EU) 2015/412 in the EU Acquis: its scope and significance

L. González Vaqué
Lawyer, Former advisor at the European Commission, China-European Food Law Working Party, Barcelona, Spain; chinaeufoodlawworkingparty@outlook.com

> I think we ought always to entertain our opinions with some measure of doubt. I shouldn't wish people dogmatically to believe any philosophy, not even mine. (Bertrand Russell)

Abstract

The scope and significance of the Directive (EU) 2015/412, and also its impact are analysed in this chapter from a strictly legal perspective. This involves, firstly, a comparative analysis of the previous regulation (contained in Directive 2001/18/EC) and that recently approved via the aforementioned Directive. Secondly, the text will address the degree of possibility of Member States being able to restrict or prohibit growth of GMOs in their territory and its compatibility with the principle of subsidiarity, and the compelling grounds on which Member States may base their restriction or prohibition. Finally, in the concluding section, there is an analysis of the new direction of GMO regulation in the EU and its (in)coherence with European Court of Justice case law, with a forward-looking perspective.

Keywords: GMO restriction or prohibition, compelling grounds, principle of subsidiarity, safeguard clauses, ECJ case law, free movement of goods

2.1 Introduction

Undoubtedly, technology related to GMOs constitutes one of the most controversial subjects, particularly in recent years (McHughen, 2013). There are many diverse reasons which explain (justify?) consumer rejection of products obtained via this technology: the difficulty in understanding exactly what it is, the lack of perception of the possible benefits that GMOs may offer (Bello Janeiro, 2015) and, even, religious beliefs and ethical considerations. In Europe in particular biotechnology applied to the agricultural sector has often been perceived in negative fashion (Marris, 2001) and the prospects for the future are increasingly pessimistic (Papatryfon *et al.*, 2008), although there are dissenting opinions (thus, for instance, the writing of this chapter coincides with the publication by various journals of the news that a group of 109 Nobel prize-winners has expressed criticism of the ecological group *Greenpeace* on account of its outright rejection of the production of GMOs, which could – in their view – represent a 'crime against humanity' (adapted from http://tinyurl.com/zhpsnjl)).

In this chapter I will attempt to analyse the scope and the impact of the legislation in question from a strictly legal perspective, avoiding taking a position with regard to the controversy between those opposed to GMOs and those in favour of their development; however, it will be impossible to explain some decisions adopted by EU institutions without referring to the negative pressure of public opinion and the diverse political opinions on the subject (Ansell *et al.*, 2006), including the influence of social media (Gunther and Liebhart, 2006).

In this context, my main objective will be to examine in some depth the provisions of Directive (EU) 2015/412 (EU, 2015) amending Directive 2001/18/EC as regards the possibility for the Member States to restrict or prohibit the cultivation of GMOs in their territory. As I have said, I will concentrate in particular upon their scope and on the elements which, according to the Community Legislator, justify the restriction or prohibition in question and their coherence in relation to European Court of Justice (ECJ) jurisprudence in this area.

2.2 A before and an after...

Prior to adoption of the aforementioned Directive (EU) 2015/412 (EU, 2015), the global legal framework for the authorisation of GMOs established by Directive 2001/18/EC (EC, 2001) and Regulation (EC) No 1829/2003 (EC, 2003b) was *fully applicable to GMOs to be used for cultivation purposes throughout the Union as seeds or other plant-propagating material ('GMOs for cultivation').* As stated in the second recital of Directive (EU) 2015/412 (EU, 2015):

> Under that legal framework, GMOs for cultivation are to undergo an individual risk assessment before being authorised to be placed on the Union market in accordance with Annex II to Directive 2001/18/EC taking into account the direct, indirect, immediate and delayed effects, as well as the cumulative long-term effects, on human health and the environment. That risk assessment provides scientific advice to inform the decision-making process and is followed by a risk management decision. The aim of that authorisation procedure is to ensure a high level of protection of human life and health, animal health and welfare, the environment and consumer interests, whilst ensuring the effective functioning of the internal market. A uniform high level of protection of health, the environment and consumers should be achieved and maintained throughout the territory of the Union. The precautionary principle should always be taken into account in the framework of Directive 2001/18/EC and its subsequent implementation.

After the entry into force of Directive (EU) 2015/412 (EU, 2015) a paragraph was added to 26 *bis* of Directive 2001/18/EC (EC, 2001) *'as from 3 April 2017 Member States in which GMOs are cultivated shall take appropriate measures in border areas of their territory with the aim of avoiding possible cross-border contamination into neighbouring Member States in which the cultivation of those GMOs is prohibited'* and two new articles the first of which is reproduced below according to the last consolidated version available:

Article 26b

Cultivation

1. During the authorisation procedure of a given GMO or during the renewal of consent/authorisation, a Member State may demand that the geographical scope of the written consent or authorisation be adjusted to the effect that all or part of the territory of that Member State is to be excluded from cultivation. That demand shall be communicated to the Commission at the latest 45 days from the date of circulation of the assessment report under Article 14(2) of this Directive, or from receiving the opinion of the European Food Safety Authority under Article 6(6) and Article 18(6) of Regulation (EC) No 1829/2003 (EC, 2003b). The Commission shall present the demand of the Member State to the notifier/applicant and to the other Member States without delay. The Commission shall make the demand publicly available by electronic means.
2. Within 30 days from the presentation by the Commission of that demand, the notifier/applicant may adjust or confirm the geographical scope of its initial notification/application.
   In the absence of confirmation, the adjustment of the geographical scope of the notification/application shall be implemented in the written consent issued under this Directive and, where applicable, the decision issued in accordance with Article 19 of this Directive as well as the decision of authorisation adopted under Articles 7 and 19 of Regulation (EC) No 1829/2003 (EC, 2003b).
   The written consent issued under this Directive and, where applicable, the decision issued in accordance with Article 19 of this Directive, as well as the decision of authorisation adopted under Articles 7 and 19 of Regulation (EC) No 1829/2003 (EC, 2003b), shall then be issued on the basis of the adjusted geographical scope of the notification/application.
   Where a demand in accordance with paragraph 1 of this Article is communicated to the Commission after the date of circulation of the assessment report under Article 14(2) of this Directive, or after receipt of the opinion of the European Food Safety Authority under Article 6(6) and Article 18(6) of Regulation (EC) No 1829/2003 (EC, 2003b), the timelines set out in Article 15 of this Directive to issue the written consent or, as the case may be, in Articles 7 and 19 of Regulation (EC) No 1829/2003 (EC, 2003b) to submit to the Committee a draft of the decision to be taken, shall be extended by a single period of 15 days regardless of the number of Member States presenting such demands.
3. Where no demand was made pursuant to paragraph 1 of this Article, or where the notifier/applicant has confirmed the geographical scope of its initial notification/application, a Member State may adopt measures restricting or prohibiting the cultivation in all or part of its territory of a GMO, or of a group of GMOs defined by crop or trait, once authorised in accordance with Part C of this Directive or with Regulation (EC) No 1829/2003 (EC, 2003b), provided that such measures are in conformity with Union law, reasoned, proportional and non-discriminatory and, in addition, are based on compelling grounds such as those related to:
   (a) environmental policy objectives;
   (b) town and country planning;
   (c) land use;
   (d) socioeconomic impacts;
   (e) avoidance of GMO presence in other products without prejudice to Article 26a;
   (f) agricultural policy objectives;
   (g) public policy.

Those grounds may be invoked individually or in combination, with the exception of the ground set out in point (g) which cannot be used individually, depending on the particular circumstances of the Member State, region or area in which those measures will apply, but shall, in no case, conflict with the environmental risk assessment carried out pursuant to this Directive or to Regulation (EC) No 1829/2003 (EC, 2003b).

4. A Member State which intends to adopt measures pursuant to paragraph 3 of this Article shall first communicate a draft of those measures and the corresponding grounds invoked to the Commission. This communication may take place before the GMO authorisation procedure under Part C of this Directive or under Regulation (EC) No 1829/2003 (EC, 2003b) has been completed. During a period of 75 days starting from the date of such communication:
   (a) the Member State concerned shall refrain from adopting and implementing those measures;
   (b) the Member State concerned shall ensure that operators refrain from planting the GMO or GMOs concerned; and
   (c) the Commission may make any comments it considers appropriate.

   On expiry of the 75-day period referred to in the first subparagraph, the Member State concerned may, for the whole duration of the consent/authorisation and as from the date of entry into force of the Union authorisation, adopt the measures either in the form originally proposed, or as amended to take account of any non-binding comments received from the Commission. Those measures shall be communicated to the Commission, the other Member States and the authorisation holder without delay.

   Member States shall make publicly available any such measure to all operators concerned, including growers.
5. Where a Member State wishes all or part of its territory to be reintegrated into the geographical scope of the consent/authorisation from which it was previously excluded pursuant to paragraph 2, it may make a request to that effect to the competent authority which issued the written consent under this Directive or to the Commission if the GMO has been authorised under Regulation (EC) No 1829/2003 (EC, 2003b). The competent authority which has issued the written consent or the Commission, as the case may be, shall amend the geographical scope of the consent or of the decision of authorisation accordingly.
6. For the purposes of an adjustment of the geographical scope of the consent/authorisation of a GMO under paragraph 5:
   (a) for a GMO which has been authorised under this Directive, the competent authority which has issued the written consent shall amend the geographical scope of the consent accordingly and inform the Commission, the Member States and the authorisation holder once this is complete;
   (b) for a GMO which has been authorised under Regulation (EC) No 1829/2003 (EC, 2003b), the Commission shall amend the decision of authorisation accordingly, without applying the procedure set out in Article 35(2) of that Regulation. The Commission shall inform the Member States and the authorisation holder accordingly.
7. Where a Member State has revoked measures taken pursuant to paragraphs 3 and 4, it shall notify the Commission and the other Member States without delay.
8. Measures adopted under this Article shall not affect the free circulation of authorised GMOs as, or in, products.

Meanwhile, Article 26c ('Transitional measures') regulates a complex and protectionist procedure of application/authorisation with broad participation of the European Commission. This provision ends with paragraph 6, which stipulates the following: *'measures adopted under this Article shall not affect the free circulation of authorised GMOs as, or in, products'.*

Furthermore, recital number 27 states that *'the provisions laid down in Articles 26b and 26c of Directive 2001/18/EC (EC, 2001) apply without prejudice to Article 23 of that Directive as well as Article 34 of Regulation (EC) No 1829/2003 (EC, 2003b)'.* The first of these provisions refers to the *safeguard clauses* and the second to the *emergency measures.*

2.3 An impregnable fortress

2.3.1 Regarding the effet utile of the new provisions

On reading the two articles to which I have just referred one deduces that their purpose is, as is indicated in the actual title of Directive (EU) 2015/412 (EU, 2015), contemplation of the possibility of Member States being able to restrict or prohibit growth of GMOs in their territory, totally or partially and for *reasons that are not strictly scientific* (Bello Janeiro, 2015) in the third section ('The Recent Reform of the Rules for GMOs Authorised for Cultivation') of the Commission communication entitled Communication from the Commission to the European Parliament, the Council, the European Economic and Social Committee and the Committee of the Regions 'Reviewing the decision-making process on genetically modified organisms (GMOs)' it recalls that *'in 2010, the Commission submitted a proposal to amend the GMO legislation to extend the grounds on which Member States could restrict or prohibit the cultivation of EU authorised GMOs on their territory ('opt-outs')'.*

Moreover, it is clear that the Community Legislator's intention was to *shield* the new *regime* against any possible challenge that might result in its annulment, bearing in mind that Directive (EU) 2015/412 (EU, 2015) was adopted due to *'the pressure exerted by countries that have opted not to cultivate GMOs in their territory'* (Carretero García, 2015a).

2.3.2 The application of the principle of subsidiarity

The sixth recital of Directive (EU) 2015/412 (EU, 2015) includes the following sentence *'experience has shown that cultivation of GMOs is an issue which is more thoroughly addressed at Member State level'.*

> Further on, in the eighth recital it is stated that: in that context, it appears appropriate to grant Member States, in accordance with the principle of subsidiarity, more *flexibility* to decide whether or not they wish to cultivate GMOs on their territory.

With regard to cross-frontier contamination, the ninth recital of the said Directive declares as follows:

> In order to ensure that the cultivation of GMOs does not result in their unintended presence in other products and whilst respecting the principle of subsidiarity, particular attention should be paid to the prevention of possible cross-border contamination from a Member State where cultivation is allowed into a neighbouring Member State where

it is prohibited, unless the Member States concerned agree that particular geographical conditions render it unnecessary.

With regard to other laws and documents, we observe that:

- Error or omission on my part excepted, the principle of subsidiarity is mentioned neither in Directive 2001/18/CE (EC, 2001) nor in Regulation n° 1829/2003 (EC, 2003b) (in the seventh recital of this Directive it is stated that '*it is necessary to approximate the laws of the Member States concerning the deliberate release into the environment of GMOs and to ensure the safe development of industrial products utilising GMOs*').
- Section 1 ('Context of the proposal') of the 'Explanatory Memorandum' of the Commission's initial proposal (document COM (2010) 375 final) recalls that, in March and June of 2009, a group of thirteen Member States (Bulgaria, Ireland, Greece, Cyprus, Latvia, Lithuania, Luxembourg, Hungary, Malta, the Netherlands, Austria, Poland and Slovenia) invited the Commission to elaborate proposals enabling Member States to decide with regard to the cultivation of GMOs. In this respect, in September 2009, the political directives of the new Commission established, at that time, by President Barroso (González Vaqué, 2010) referred to the principle of subsidiarity in the area of GMOs as an example of how it was not always possible to for there to be a balance between the framework of the EU and the need to take into account diversity in a Union of 27 Member States: '*according to these guidelines, it should be possible to combine a European Union authorisation system for GMOs, based on science, with freedom for Member States to decide whether or not they wish to cultivate GM crops on their territory*' (see also Section 1 of the 'Explanatory Memorandum' of the aforementioned document (COM)2010 375 (EC, 2010)).

To complete this section I feel it is appropriate to recall that the Community Legislator, in Directive (EU) 2015/412 (EU, 2015), uses on several occasions the term/concept *flexibility*, such as, for example:

- In the sixth recital: '*issues related to the placing on the market and the import of GMOs should remain regulated at Union level to preserve the internal market (but, the cultivation may however require more flexibility in certain instances as it is an issue with strong national, regional and local dimensions, given its link to land use, to local agricultural structures and to the protection or maintenance of habitats, ecosystems and landscapes (of the Member States)).*'
- In the eighth recital, the text of which I have already referred to, subsidiarity and *flexibility* appear to be consecutively interrelated.

In addition, the aforementioned third section of document COM(2015) 176 (EC, 2015), reiterates that '*the 2015 Directive...gives Member States more flexibility to decide whether or not they wish to cultivate GMOs on their territory, whilst still maintaining the system of EU authorisation based on risk assessment.*'

2.3.3 New compelling grounds (not strictly scientific)

I think it appropriate to indicate that the modification of Directive 2001/18/EU (EC, 2001) initially proposed by the Commission (document (COM) 2010 375 final (EC, 2010), mentioned above) was not very detailed and, probably, lesser in scope, as it confined itself to providing for the insertion in this Directive of the following article:

Article 26b

Cultivation

Member States may adopt measures restricting or prohibiting the cultivation of all or particular GMOs authorised in accordance with Part C of this Directive or Regulation (EC) No 1829/2003 (EC, 2003b), and consisting of genetically modified varieties placed on the market in accordance with relevant EU legislation on the marketing of seed and plant propagating material, in all or part of their territory, provided that:

(a) those measures are based on grounds other than those related to the assessment of the adverse effect on health and environment which might arise from the deliberate release or the placing on the market of GMOs;
and,
(b) that they are in conformity with the Treaties.

By way of derogation to Directive 98/34/EC, Member States that intend to adopt reasoned measures under this Article shall communicate them to the other Member States and to the Commission, one month prior to their adoption for information purposes.

On the other hand, the modification introduced by the 2015 Directive includes a whole battery of *compelling grounds* (in the French version '*motifs sérieux*') with which I deal shortly. Indeed, according to the third point of the new article 26b, national measures to restrict or prohibit the cultivation of GMOs authorised at EU level must be *in conformity with Union law, reasoned, proportional and non-discriminatory* and, moreover, must be based upon *compelling grounds such as those related to*:

(1) Environmental policy objectives: I believe that these will be scientific arguments different from those taken into account in the individual assessment of the risks associated with the GMO in question, prior to authorisation of their commercialisation on the Union market according to annex II of Directive 2001/18/EC (EC, 2001), bearing in mind any direct, indirect, immediate, delayed or long-term cumulative effect on human health and the *environment* (and, in particular, will refer to the environmental policy of the Member State that invokes this motive).
(2) Town and country planning: this is a *compelling ground* that leaves considerable room for manoeuvre to Member States and, if we take into consideration what is indicated in recital number 15 of Directive 2015/142 (EU, 2015), town and country planning is included in the category of *other* grounds upon which the States in question can base their adoption of measures insofar as they are '*other legitimate factors including those relating to cultural traditions (traditions that are not mentioned again per se in this EU regulation)*'.
(3) Land use: my comments regarding the previous *compelling ground*, is equally valid in this case given that, in recital number 15 of Directive 2015/142 (EU, 2015), which I have just mentioned, it is also included in the category of legitimate factors to be taken into account (as I have already said, in the sixth recital of the Directive in question, it is stated that, although questions related to commercialisation and import of GMOs must continue to be regulated at Union level, '*cultivation may however require more flexibility in certain instances as it is an issue with strong national, regional and local dimensions, given its link to land use*'.
(4) Socioeconomic impacts: this is one of the most controversial *compelling grounds* because the ECJ has repeatedly opposed justification for economic reasons of obstacles to free circulation of goods

and a broad interpretation of these impacts would represent a serious threat to the functioning of the domestic market.

(5) Avoidance of GMO presence in other products without prejudice to Article 26a: the community legislator considers that Member States should also have the possibility of adopting measures restricting or prohibiting the cultivation of GMOs authorised in all or part of their territory in accordance with the Directive and that '*those grounds may be related to the high cost, impracticability or impossibility of implementing coexistence measures due to specific geographical conditions, such as small islands or mountain zones, or the need to avoid GMO presence in other products such as specific or particular products*' (see also recital number 15 of Directive 2015/412 (EU, 2015).

(6) Agricultural policy objectives: the Community Legislator, in the oft-cited recital number 15 of Directive 2015/412 (EU, 2015), referrers to the information provided by the Commission, as was requested in the conclusions of the Council of 2008, on the socioeconomic impact of GMO cultivation and underlines the fact that that '*the outcome of that report may provide valuable information for Member States considering taking decisions on the basis of this Directive [with regard to] grounds relating to agricultural policy*.'

(7) Public policy (in the French version '*l'ordre public*') is a reason that should be interpreted with caution and proportionality so as not to (in)directly support a prohibition of GMO cultivation giving in to the violence exercised by certain groups of activists (Hayes, 2007) and, in my opinion it is necessary to refer to the case law enshrined in, amongst others, the 'Fresas' (case C-265/95) and 'Brenner Motorway' (case C-112/00) ECJ judgements and interpret/apply them *mutatis mutandis* case by case.

In conclusion, as is confirmed in the aforementioned third paragraph of document COM(2015) 176 final (EC, 2015) it is a question of '*compelling reasons other than the risk to human or animal health and the environment that is, criteria other than those assessed by European Food Safety Authority (EFSA) in its risk assessment*' and I think the Community Legislator was correct in replacing the ambiguous expression '*based on grounds other than those related to the assessment of the adverse effect on health and environment*' proposed by the Commission, with a heterogeneous but, insofar as is possible, fairly specific list of grounds, which guarantees the principle of legal certainty.

Two final observations regarding this list:

- these are compelling grounds such as those *related to...* for which reason one cannot exclude the possibility of the ECJ having to interpret and define the scope of this 'relationship'; and
- also in the light of the said expression, it is worth considering whether the list in question should be regarded as exhaustive... in any case, if we take into account that it lists *exceptions* to a general rule, in principle, this list should be interpreted restrictively.

If anyone laments the absence in the so oft-mentioned list of compelling grounds of a reference to ethics, they should bear in mind that the ninth recital of Directive 2001/18 emphasises the importance of respect for the ethical principles recognised in each Member State; and in its article 29 regulates the 'Consultation of Committee(s) on Ethics' '*without prejudice to the competence of Member States as regards ethical issues*'.

2.3.4 The Free Movement of goods

It comes as no surprise that the Community Legislator has taken pains to explain that the new articles included in Directive 2001/18 (EC, 2001) are not incompatible with the provisions of the Treaty concerning the free movement of goods, and that they cannot be interpreted/applied in such a manner that they violate those provisions.

In this respect, the fifth recital of Directive 2015/412 (EU, 2015) confirms that, in principle, once a GMO has been authorised for cultivation, in accordance with prevailing EU legislation, and it complies with, in terms of the type that is going to be commercialised, the requirements of EU Law regarding the commercialisation of seeds and other plant propagating material, Member States are not authorised to prohibit, restrict or prevent its free movement in their territory... although, in seemingly redundant (superfluous?) fashion, the end of this recital, specifies '*except under the conditions defined by Union law*'.

I shall emphasise that paragraph 6 of the new article 26b of Directive 2001/18/EU (EC, 2001), reiterates that '*measures adopted under this Article shall not affect the free circulation of authorised GMOs as, or in, products*', without further specification. This provision corresponds to what is stated in recital number 16 of Directive 2015/412 (EU, 2015): '*the restrictions or prohibitions adopted pursuant to this Directive should refer to the cultivation, and not to the free circulation and import, of genetically modified seeds and plant propagating material as, or in, products and of the products of their harvest*'; and, obviously, '*[they] should, furthermore, be in conformity with the Treaties, in particular as regards the principle of non-discrimination between national and non-national products, the principle of proportionality and Article 34, Article 36 and Article 216(2) Treaty on the Functioning of the European Union (TFEU)*' (*ibidem*).

There is no doubt whatsoever that, in any case, *the free movement and import of genetically modified seeds and plant propagating material* is guaranteed, but... what about their *use*? We have to refer in this respect to ECJ case law in two complex cases (C-110/05 ('Commission/Italy', of February 10, 2009) and C-142/05 ('Mickelsson and Roos', of June 4, 2009)): briefly, we recall that the origin of this *entanglement* was an unusual proposal by Advocate General Kokott, formulated in his Opinion regarding the second of these cases, which proposed the widening of the scope of the case law 'Keck and Mithouard' (joined cases C-267/91 and C-268/91) to the *forms of use*; in other words, according to Advocate General Kokott, just as the selling arrangements do not constitute an obstacle to the free movement of goods, neither would what she called 'arrangements for use'. This peculiar proposal revived the debate regarding the use and efficiency of a formalistic and inefficient method to determine the applicability of the article in the Treaty which prevents measures from having an equivalent effect to that of quantitative restrictions on imports (currently article 34 TFEU). Finally, the ECJ requested a *new* Opinion from Advocate General Bot: from these Opinions and the sentences passed in the two aforementioned cases one deduces that the provisions of the Treaty regarding the free movement of goods are wholly applicable to *arrangements for use* and, in particular the prohibition of use or utilisation (if memory serves me right, one of the most convincing arguments referred to food additives: what is the point of authorising the import and sale of an additive, if its use is prohibited?).

It will certainly be necessary to invoke case law (not only of the ECJ, but of many Member States) which justifies *inverted discrimination*, based on the French concept of '*discrimination à rebours*', which tends, according to various modalities, to give a considerable advantage to similar economic operators from other Member States accepting that this might prevent, directly or indirectly, effectively or potentially, exchanges between Member States.

2.3.5 Safeguard clauses and emergency measures

Although, in the past, to restrict or prohibit GMO cultivation, some Member States invoked safeguard clauses and emergency measures according to article 23 of Directive 2001/18/EC (EC, 2001) and article 34 of Regulation (EU) n° 1829/2003 (EC, 2003b) (seventh recital of Directive 2015/412 (EU, 2015), in my opinion the considerable extension of the *compelling grounds* in order to request the territorial restriction or prohibition of GMO cultivation will mean that, in the future, these measures will seldom be applied.

2.4 Conclusions

2.4.1 The new orientation and its in(coherence) with European Court of Justice case law

I cannot neglect to mention that the ECJ has passed some judgments rejecting the adoption of national measures prohibiting the use of GMOs in certain territories. In this respect, I shall refer to the paradigmatic ruling 'Austria/Commission', of September 13, 2007, joined cases C-439/05 P and C-454/05 P.

Via the judgement in question, the ECJ decided to dismiss the appeals lodged by the *Land Oberösterreich* and the Republic of Austria requesting the annulment of the ruling by the the Court of First Instance of the European Communities, of October 5, 2005 (joined cases T-366/03 and T-235/04), on a draft bill, notified to the Commission in 2003: the *Oberösterreichisches Gentechnik-Verbotsgesetz de 2002*, which prohibited *genetic engineering*. The aim of this project was to prevent the cultivation of seeds and plants composed of or containing GMOs, as well as the breeding and releasing of transgenic animals for the purpose of hunting and fishing (the aim of the notification was to obtain, on the basis of article 95 EU, Section 5, an exception to the provisions of Directive 2001/18 (EC, 2001).

Bevitatis causae, I shall now list just some of the arguments employed by the ECJ in its decision not to annul Commission Ruling 2003/653/EU, of September 2, 2003, in particular:

- Pointing out that the legality of the national measures notified according to article 95 EU, Section 5, were closely linked to appreciation of the scientific innovations provided by the Member State which presented a notification (in effect, the prevailing legislation at that time required that the adoption of national provisions establishing exceptions to a harmonising measure be based on scientific innovations related to the protection of the workplace or the environment and justified by a specific problem of the Member State in question that had arisen subsequent to the adoption of the harmonising measure, and that the proposed provisions, and the grounds for their adoption be notified to the Commission – see paragraph 80 of judgement 'Germany/

Commission', of January 21, 2003, case C-288/96) (paragraphs 56 and 57 of the judgement 'Austria/Commission').

- Acknowledging that the Court of First Instance had rightly concluded that the appellants had not provided data invalidating the conclusions of the EFSA, according to which the Republic of Austria did not demonstrate that the territory of *Land Oberösterreich* possessed exceptional or unique ecosystems which necessitated a risk assessment different to that performed with regard to Austria as a whole or to other similar regions in Europe (*ibidem*, paragraph 61).
- In confirming the validity of the EFSA judgement, which stated the lack of scientific innovation demonstrating the existence of a specific problem (see the results of the said EFSA judgement) (*ibidem*, paragraph 63).
- As well as denying that the Court of First Instance committed an error of Law in considering that the appellants had not provided elements of proof that called into question the basis of the findings regarding the lack of scientific innovation demonstrating the existence of a specific problem and that, therefore, there had been non-compliance with one of the requirements established in article 95 EU, paragraph 5 (*ibidem*, paragraph 68).

The question arises as to whether judgement on a similar or comparable case referring to Directive 2001/18 (modified by Directive 2015/412 (EU, 2015)) would be the same: the answer is surely a negative one, given that the grounds that justify Member States restricting or prohibiting GMO cultivation in their territory are no longer exclusively based on a risk assessment of direct, indirect, immediate and delayed effects, as well as the cumulative long-term effects, on human health and the environment of the GMO in question.

2.4.2 The Commission's participation in new authorisation procedures

I shall not dwell on an analysis of the characteristics of the EU reporting procedure applied to measures adopted by Member States in accordance with the new version of Directive 2001/18 (EC, 2001). But I will highlight the significance of the Commission's intervention, time and time again, in the various stages of notification, authorisation, etc. provided for in the new article 26c.

With regard to transparency, recital number 20 of Directive 2015/412 (EU, 2015) indicates that:

- '*Given the importance of scientific evidence in taking decisions on the prohibition or approval of GMOs, the Authority should collect and analyse the results of research regarding the risk or danger to human health or the environment of GMOs and inform the risk managers of any emerging risks.*'
- '*Such information should be made available to the public.*'

In this regard, at the end of paragraph 1 of said article 26b it is established that the Commission will make the requests referred to in this paragraph '*publicly available by electronic means*' (see also the tenth recital and articles 9 and 24 of Directive 2001/18 (EC, 2001)).

And I believe that these provisions should be based upon the logic of the 'Principles of Transparency' included in Section 2 of Regulation (EC) No 178/2002 of the European Parliament and of the Council of 28 January 2002 laying down the general principles and requirements of food law, establishing the European Food Safety Authority and laying down procedures in matters of food safety (OJ L 31, 1.2.2002, p. 1); this is not a case of organising voting sessions based on a show of hands in favour of or against reliable scientific assessments provided by the relevant agencies, or of conducting reverse auctions to appease certain populist movements or political patronage.

2.4.3 Prospects for the future

It is true that, as scholars have indicated (Carretero García, 2015b) the Community Legislator has yet to address the provision of solutions to some important problems; for example, the one that is mentioned, *inter alia*, in the Opinion of the European Economic and Social Committee on the 'Proposal for a Regulation of the European Parliament and of the Council amending Directive 2001/18/EC (EC, 2003a)' (which, eventually, was adopted as a Directive), in other words, '*the need for EU-level minimum norms for coexistence and liability*'.

In any case, Directive 2015/412 (EU, 2015) establishes that:

- by April 3, 2019 at the latest, the Commission will present to the European Parliament and to the Council a report on the use made by Member States of this Directive, including the efficiency of the provisions permitting Member States to restrict or prohibit GMO cultivation in all or part of their territory and the correct functioning of the domestic market and, furthermore, that report will be accompanied by the legislative proposals deemed appropriate by the Commission (article 2); and
- by April 3, 2019 at the latest, the Commission will update the annexes of Directive 2001/18/EU (EC, 2001) in accordance with article 27 of said Directive with regard to the assessment of risk to the environment, in order to include and develop the reinforced guidance of the EFSA of 2010 on environmental risk assessment of genetically modified plants (article 3).

Therefore, in the mid-term there will be objective evaluation of the results of the provisions introduced in Directive 2001/18/EU (EC, 2001) in accordance with the so oft-cited Directive 2015/412 (EU, 2015).

References

Ansell, C., Maxwell, R. and Socurelli, D. (2006). Protesting foods: NGOS and political mobilization in Europe. In: Ansell, C. and Vogel, D. (eds.) Politics, science, and the environment: what's the beef? The contested governance of European food safety. MIT Press, Cambridge, MA, USA, pp. 97-119.

Bello Janeiro, D. (2015). El futuro legal en materia de transgénicos. Revista de Derecho Agrario y Alimentario 67: 33-35.

Carretero García, A. (2015a). La Unión Europea permite a través de la Directiva 2015/412 que los Estados restrinjan o prohíban el cultivo de organismos modificados genéticamente en su territorio. Revista de Derecho Agrario y Alimentario 67: 40-42.

Carretero García, A. (2015b). Coexistencia, etiquetado, control y restricción o prohibición del cultivo de organismos modificados genéticamente en la Unión Europea. Revista CESCO de Derecho de Consumo 14: 201-202.

European Commission (EC) (2001). Directive 2001/18/EC of the European Parliament and of the Council of 12 March 2001 on the deliberate release into the environment of genetically modified organisms and repealing Council Directive 90/220/EEC. Official Journal of the European Union L 106: 1.

European Commission (EC) (2003a). Proposal for a Regulation of the European Parliament and of the Council amending Directive 2001/18/EC as regards the possibility for the Member States to restrict or prohibit the cultivation of GMOs in their territory – document COM(2010) 375 final. EC, Brussels, Belgium.

European Commission (EC) (2003b). Regulation (EC) No 1829/2003 of the European Parliament and of the Council of 22 September 2003 on genetically modified food and feed. Official Journal of the European Union L 268: 1.

European Commission (EC) (2010). Opinion of the European Economic and Social Committee on the 'Proposal for a Regulation of the European Parliament and of the Council amending Directive 2001/18/EC as regards the possibility for the Member States to restrict or prohibit the cultivation of GMOs in their territory' COM(2010) 375 final. Official Journal of the European Union C 54: 51.

European Commission (EC) (2015). Communication from the Commission to the European Parliament, the Council, the European Economic and Social Committee and the Committee of the Regions 'Reviewing the decision-making process on genetically modified organisms (GMOs)' – document COM(2015) 176 final. EC, Brussels, Belgium.

European Union (EU) (2015). Directive (EU) 2015/412 of the European Parliament and of the Council of 11 March 2015 amending Directive 2001/18/EC as regards the possibility for the Member States to restrict or prohibit the cultivation of genetically modified organisms (GMOs) in their territory. Official Journal of the European Union L 68: 1.

González Vaqué, L. (2010). El Derecho de la UE relativo a los organismos modificados genéticamente: la Comisión Europea cambia de estrategia para permitir, restringir o prohibir su cultivo'. Revista de Derecho y Genoma Humano 33: 217-240.

Gunther, A.C. and Liebhart, J.C. (2006). Broad reach or biased source? Decomposing the hostile media effect. Journal of Communication 56: 449-466.

Hayes, G. (2007). Collective action and civil disobedience: the anti-GMO campaign of the Faucheurs Volontaires. French Politics 5: 293-314.

Marris, C. (2001). Public views on GMOs: deconstructing the myths – Stakeholders in the GMO debate often describe public opinion as irrational. But do they really understand the public? EMBO Reports 2: 545-548.

McHughen, A. (2013). GM crops and foods: what do consumers want to know? GM Crops and Food 4: 172-182.

Papatryfon, I., Zika, E., Wolf, O., Gómez-Barbero, M., Stein, A.J. and Bock, A.-K. (2008). Consequences, opportunities and challenges of modern biotechnology for Europe – The analysis report. European Commission, Joint Research Centre, EUR 23413 EN – 2007, 240 pp.

3. Restrictions on GM crops in the European Union: between legislative unity and the diversity of national positions

F.J. Romero-Caro
Associate Researcher, Department of Constitutional Law and History of Political Thought, University of the Basque Country, P.O. Box 644, 48080 Bilbao, Spain; franciscojavier.romero@ehu.eus

Abstract

Directive (EU) 2015/412 amended Directive 2001/18/EC as regards the possibility for Member States (MS) to restrict or prohibit the cultivation of genetically modified organisms (GMOs) in their territory. This chapter analyses the opt-outs implemented by MS under the aegis of the Directive since it came into force. It starts by setting out in detail the mechanisms provided by that Directive for any MS so wishing to restrict or prohibit, in all or part of their territory, GMOs crops that have been approved at EU level in accordance with the current legislation. It then describes the way in which 19 of the 28 MS have made use of this opt-out prerogative, either totally or partially, with particular emphasis on compound states (those with federal or similar structures) where the restriction has been applied only with regard to one part of the territory. Focusing primarily on opt-out states, the study also analyses another key element: the GMOs that are targeted by the restriction. The analysis identifies a common pattern among opt-out MS whereby they declare their territories to be 'GMO-free'. This chapter also looks at the formal aspect of notification and the way in which opt-out MS have justified the restrictions applied to GMOs in their territory, within the compelling grounds listed in the Directive. With regard to formal aspects, it is noted, *inter alia*, how opt-out MS have failed to meet the deadline for submitting notification and certain other details relating to MS in which the exclusion on GMOs only affects part of the national territory. As for the grounds given for the restrictions, it examines the way in which most opt-out states have circumvented the requirement to justify their restrictions by resorting to a legal device. It also analyses the vagueness of the arguments put forward by Germany – the only State to officially set out grounds for restriction. In order to attempt to understand this lack of transparency in the exercise of their opt-out faculties, the chapter finally examines other instruments to try to find the grounds on which 19 MS may have decided to place restrictions on GM crops that had been approved at EU level following rigorous assessment of the environmental and human health risks.

Keywords: mechanisms to request restriction of cultivation, territorial exclusions, restricted crops, notification of the opt-out

3.1 Analysis of the way in which Member States are making use of the powers conferred upon them by Directive 412/2015: introductory questions

Since its beginnings, genetic engineering has been a politically controversial subject, particularly in the farming and food sectors. This controversy has generated numerous currents of opinion, both for and against, sensitising states to the issue and determining their modus operandi in this area. As other authors of this monograph have noted (López Basaguren), the decision-making process with regard to genetically modified organisms (GMOs) in the sphere of the European is more complex given that it involves both environmental and agricultural competences which, especially the latter, have been largely ceded by the Member States (MS) in accordance with articles 39 to 44 of the Treaty on the Functioning of the European Union (TFEU). The problems experienced with the initial legal framework, implemented in the 1990s, led the EU to articulate the framework currently in force, which is based upon Directive 2001/18/EC (EC, 2001) and Regulations 1829 and 1830/2003 (EC, 2003a,b). This is a regulatory framework structured on two levels for the authorisation of cultivation and for use for food or feeds, which underwent some reforms as a consequence of Directive 412/2015 (EU, 2015). The first tier, of a technical nature, decision-making with regard to GMOs for agri-food use in the EU involves a scientific-technical risk assessment of the genetically modified organism in question by the EFSA (European Food Safety Authority), which will issue a technical opinion in this regard. In the case of GMOs intended for cultivation, the requested member state will have to perform its own assessment before submitting the request to the EFSA. The second tier, of a clearly political nature, implies acceptance or rejection of the GMO request be it for cultivation or for their use in food and feed. In this phase there is a process of codecision between the European Commission (EC) and the MS, which express their opinion via a Committee of Experts. The EC presents a draft resolution, and the MS decide on approval or rejection by qualified majority. In the event of not reaching a decision in either sense a non-opinion is issued. In the latter case, the EC may again present a draft before the Appeal Committee, which decides in the same way as the Committee of Experts. In the event of continued non-opinion, the European Commission may take a unilateral decision with regard to adoption of the GMO in question.

This final unilateral decision by the European Commission in cases in which there is no agreement between the MS may result in considerable conflict within the Union owing to the special political significance of the question. In fact, various ME – among others, Austria, Germany, Hungary or Poland – have invoked the safeguard clause and emergency measures in order to avoid these authorisations having an impact in their territory. Italy invoked the precautionary principle to restrict GMO cultivation in its territory. The case of France is more significant as it invoked an emergency measure to ban the cultivation of MON 810. This decision was annulled by the European Court of Justice (ECJ)[26] in its ruling of September 8, 2011 interpreting that the French State had erred in invoking the legal principle via which it articulated this measure (paragraph 63) in doing so on the basis of Directive 90/220/EEC (EU, 1990) of the Council, of April 23, 1990 and not Regulation n° 1829/2003 (EC, 2003a). Therefore, the ECJ deemed legitimate the intention to restrict cultivation, though it was necessary for this decision to be based upon the appropriate regulatory instrument (ECJ, 2013).

With the aim of avoiding these conflicts, which have been addressed by another contributors to this volume (López Basaguren; Escajedo San-Epifanio) and given the evident diversity of opinions among MS, Directive 2015/412 of March 11 was adopted to allow MS who so wished to restrict or prohibit GMO cultivation in their territory.

[26] Court of Justice Ruling of September 8, 2011 related to issues C-58/10 and C-68/10.

This Directive permits MS to exercise a right to opt-out of the cultivation of the GMO in question. The raison d'être of this measure is specified in Consideration 6, which shows that cultivation is a subject analysed in greater depth by the MS given its significant national, regional and local dimensions which necessitates greater flexibility in the uniform application of the authorisation.

This work compiles the results of a study of the way in which 19 of the 28 MS have exercised the opt-out option offered by Directive 412/2015 (EU, 2015). It explains the mechanisms that the Directive provides to apply for total or partial restriction of cultivation, and the territorial exclusions notified to date, indicating the territorial extension of the latter and the crops that have been the subject of restriction. Attention is also paid to the notification process and the grounds that, where appropriate, different MS have cited to justify their decision.

3.2 The mechanisms to request restriction of cultivation

Directive 2015/412 (EU, 2015) grants MS the possibility of imposing restrictions, either partial or total, on the cultivation of GMOs in their territory. To this end, the Directive provides two different mechanisms depending on the moment when the communication is produced. Firstly, in light of article 1 of the aforementioned Directive – which inserts article 26b into Directive 2001/18/CE – it is possible to request adaptation of the geographical scope of application of the authorisation to restrict or exclude a territory from cultivation during the authorisation procedure of a particular GMO or during the extension of an authorisation already in force. The Commission's communication must be within 45 days of issue of the assessment report. As will be discussed below, this has been the route employed by MS, which was already provided for by community instructions as expressed in Consideration 19 of the aforementioned Directive.

Furthermore, Section 3 of article 26b permits the adoption of measures to restrict or prohibit the cultivation of a GMO in the event of absence of the declaration contemplated in Section 1 or if the declaration is made to restrict the cultivation of an already authorised GMO. Article 3 of the Directive requires that these measures be in conformity with Union Law, reasoned, proportional and non-discriminatory and, in addition, based on a series of grounds. These are related to: (1) environmental policy objectives; (2) town and country planning; (3) land use; (4) socioeconomic impacts; (5) avoidance of GMO presence in other products; (6) agricultural policy objectives and (7) public policy.

Those grounds may be invoked individually or in combination, with the exception of public policy, which must be invoked in combination with one of the other grounds. The Directive does not provide a definition of these concepts, so interpretation of the latter corresponds to both the MS and the Commission. This provision seems to invoke undefined legal concepts in order to offer states greater flexibility when it comes to imposing restrictions on cultivation. The decisive interpretation will be that adopted by the Commission, as it is the organism responsible for analysing and validating declarations made by states. Ultimately, the ECJ may have to rule on the validity of a hypothetical declaration.

Directive 2015/412 (EU, 2015) grants MS – via the creation of article 26c – a transition period, which ended on October 3, 2015, during which to notify the adaptation of the geographical scope of the cultivation applications for those submitted and/or approved before April 2, 2015. Pursuant to this law, MS wishing to exclude their territory from the cultivation of authorised GMOs or whose assessment report by the EFSA has been presented must notify their intention before the date indicated. Upon expiry of the deadline, the Member State wishing to exclude these GMOs must invoke the procedure explained previously in Section 3 of article 26b. In the event of the EFSA report being pending, the MS have 45 days following publication of the latter in which to notify restrictions as regards the geographical scope of cultivation.

Some MS have issued reports on the socioeconomic impact of GMOs, which they may take into account when interpreting these criteria. These include those of Austria (Spöck, 2010) (Greiter *et al.*, 2011), and the Netherlands (COGEM, 2009). Scientific literature has also expressed views in this regard, without there being consensus regarding these categories (Fischer *et al.*, 2015). Therefore, comparative analyses like Apolline's are useful when considering the divergences between states (Apolline, 2015). With the goal of guaranteeing greater legal security it would help if the Commission were to clarify how the concepts contained in the directive should be understood in the same way as it does in article 2 of Directive 2001/18/EC. It would also be a positive step to detail which supposed practices are included in each of these concepts.

Of the 28 MS, 17 notified on time the adaptation of the geographical scope of the authorisations to restrict cultivation throughout their territory, 2 partially excluded their territory, whilst 9 have not issued a notification to date.

On the basis of the above it is clear that the Directive responded to the demands of the MS given that most of them have used the mechanism of article 23b within the transition period. These MS have favoured the implementation of some kind of restriction on GMO cultivation in their territory, the majority of them opting for total exclusion of cultivation of all authorised GMOs or those in the process of being authorised. However, some states have opted for partial exclusions either of their territory or among the excludable GMOs. The form and justification of notification is another aspect that has differed between MS.

3.3 Territorial exclusions notified to date

Directive 2015/412 (EU, 2015) permits MS to effect total or partial exclusion of cultivation of GMOs in their territory. This measure grants flexibility to MS, allowing for the coexistence of GMO-free zones with zones where their cultivation is permitted within the same state. The possibility of implementing a partial exclusion is very closely linked to what is expressed in consideration 6, which acknowledges the strong local and regional division of cultivation. Partial exclusions are a modality that responds perfectly to the needs of compound states. This mechanism enables states to accommodate different sensibilities with regard to this question coexisting amongst the subnational units that integrate them.

To date, 17 of the MS have given notification of the total exclusion from their territory of one or more GMO. These states are: Austria, Bulgaria, Croatia, Cyprus, Denmark, France, Germany, Greece, Hungary, Italy, Latvia, Lithuania, Luxembourg, Malta, the Netherlands, Poland and Slovenia.

3.3.1 Partial territorial exclusions

Meanwhile, both Belgium and the UK have notified the European Commission of their desire to exclude cultivation from part of their territory. In the case of Belgium, the exclusion of cultivation only affects the Walloon Region, with cultivation possible in both Flanders and the Region of Brussels-Capital. In Flanders it is obligatory to notify both the regional authority and the owners of neighbouring farms of the intention to cultivate a GMO (Flemish Government, 2010). The reason for this difference between territories is based on the distribution of competences produced within the Belgian federation. Following the reform of the Belgian state in 2014, competence in agricultural and environmental areas belongs to the regions, so they decide whether they wish to declare their territory free of GMO cultivation. The Walloon Region cannot unilaterally restrict this cultivation since as it is part of a MS of the European Union it is the latter that must issue the notification. The Belgian federation notified the Commission of the Walloon Region's desire by means of a Cover Letter, in which it communicated the will of the Walloon Ministry of Agriculture and attached the pertinent notifications. It also communicated the fact that neither of the other two regions that integrate the federation had expressed a will to issue a similar communication. Despite the fact that the federal government – via the Health Department –, acted as a mere intermediary transmitting the will of one of its sub-state entities, it is necessary to point out that without its acquiescence it would not have been possible to exclude cultivation in the Walloon Region. Although at an internal level competence in the area of agricultural cultivation belongs to the region, the community Directive grants the possibility of restricting or prohibiting GMO cultivation exclusively to MS irrespective of their internal organisation. Therefore, only a Member State can perform such an exclusion, so that without its active participation communicating the request to the Commission the exclusion in question cannot be implemented.

The other MS that has effected a partial exclusion of its territory is the UK. The British government sent the Commission a series of Cover Letters in which it requested the exclusion of the territory of Scotland, Wales and Northern Ireland from GMO cultivation. Thus, GMO cultivation is only permitted in England. The territory of Gibraltar, which forms part of the EU though it is excluded from the Common Agricultural Policy, has no agriculture so the exclusion, or not, of the territory has no practical consequences.

As in the Belgian case, the UK government communicated to the Commission the exclusion of GMO cultivation in the territory of these three nations at the request of their administrations. The UK is a unitary state in which there exists a system of devolution of power. The Parliament of Westminster has devolved competences to the Parliaments of these three territories. In the case of Scotland the *Scotland Act* (GS, 1998) follows a system of negative delimitation of competences. According to this, those areas that are not reserved to the British government under Annex 5 will be competence of the Parliament of Hollyrood. Agriculture and environment are not included in these, so competence in these fields resides with Scotland. The *Government of Wales Act 2006* (GW, 2006), in Annex 5, assigns to the Welsh Assembly competence in agriculture (Section 1) and environment (Section 6). In the case of Northern Ireland, all those competences are devolved which are not included in Annex 2 (*excepted matters*) and 3 (*reserved matters*) of the *Northern Ireland Act 1998* (GNI, 1998) later modified in 2006.

As mentioned earlier, the mere will of the government of any of these territories to exclude cultivation is not sufficient, as it is the Member State that is responsible for notifying the Commission of the decision. Furthermore, the British Parliament by virtue of its parliamentary sovereignty could have disallowed these petitions and not authorised the government to issue these notifications of exclusion of cultivation. Given that the British government has not considered it opportune to implement any restriction in the territory of England, inside the UK there co-exists a GMO-free zone with another where their cultivation is permitted.

Germany excludes its entire territory though, Federal state that it is, it issues the communication detailing each Federated state as it is the latter that possess competence in agricultural matters. This form of communicating exclusion appears to indicate that in the event of a land wishing to remove the restriction from its territory the federal government would communicate this decision to the European Commission.

The 9 remaining MS have not introduced any type of exclusion, so GMO cultivation is authorised throughout their territory. These states are: the Czech Republic, Estonia, Finland, Ireland, Portugal, Romania, Slovakia, Spain, and Sweden.

Within these MS there are some zones that have attempted to exclude GMO cultivation in their territory. In the case of Spain, on June 18, 2015 the Basque Parliament approved a non-legislative motion expressing its desire to declare the three provinces of the Basque Autonomous Community free of GMO cultivation, and requesting from the government of the nation competence to take decisions in this area (BG, 2015). The request issued by the Basque Parliament is unviable, as it is the government of the MS which has competence to notify of the adaptation of the territorial scope of cultivation. It can only request that central government notify the exclusion, as the latter can decide independently with regard to the question, given that although competence in agriculture is partly transferred (148.7 Spanish Constitution) the bases and coordination of the general planning of the economy, and relations with the EU are exclusive competences of the central State (149.13 and 149.3 Spanish Constitution respectively). This type of request from the Basque Country is not new, as in 2007 the Basque Government established its position in this matter declaring the 'the Autonomous Community of the Basque Country an area free of transgenic crops intended for human consumption or animal feed' (BG, 2007).

Other Autonomous Communities like Asturias, the Balearic and the Canary Islands have in the past adopted resolutions declaring their territory free of GMO cultivation, though these have only a declaratory value given that these Communities lack competence to prohibit this cultivation. These resolutions are merely political declarations that have no legal value whatsoever. Moreover, generally speaking, they are issued in places where there is no GMO cultivation, reducing their impact further still. Empirical evidence shows that in the aforementioned territories there is no GMO cultivation.

To date the Spanish Government has not deemed it necessary to issue any notification to the European Commission so there is currently no restriction on GMO cultivation on Spanish territory.

Portugal has not communicated any restriction on cultivation on the basis of the existing Directive. However, in 2010 it issued a request based on former article 95.5 of the European Community

Treaty (ECT) – now article 114.5 of the TFEU – requesting the prohibition of GMO cultivation in the ultra-peripheral region of Madeira as a necessary measure to protect the environment. The Commission referred the request to the EFSA in order that it might issue an expert opinion on the question. The EFSA analysed the information supplied by Portugal and concluded that there existed no new evidence to suggest risks to human and/or animal health or to the environment justifying the prohibition sought by Portugal (EFSA, 2010). Despite the fact that the EFSA found no reason to justify the Portuguese request, the European Commission waited for expiration of the six-month deadline available in which to establish its position, in such a way that the request made by Portugal is deemed to be approved according to article 95.6 TCE – now 114.6 TFEU.

In Finland, Romania and Sweden some municipalities have passed motions in this respect, albeit they are of merely declaratory value and in no way legally binding for their governments. Ireland, which in 2009 announced its intention not to permit the cultivation of GMOs in its territory and which actively supported Directive 2015/412 (EU, 2015), allowed the deadline to notify a restriction of prohibition to expire. These changes are due to the fact that the 2009 declaration – which did not become law – occurred during negotiations to form a coalition government that required the support of the Green Party. The subsequent government opted for a different policy taking into account Irish cattle's high dependency upon genetically modified feed.

3.4 Restricted crops

The MS can restrict the cultivation of those GMOs that have been authorised by the Commission, or whose authorisation is being processed. Table 3.1 shows the GMOs authorised or in the process of being so for cultivation in the European Union.

As we can see, currently only one GMO, Maize MON 810, has been approved for cultivation in the European Union. This GMO is not the only one that has been approved for this purpose, since in 2010 the Commission approved cultivation of the potato for industrial use EH92-527-1, also

Table 3.1 Genetically modified organisms authorised of in process of being so for cultivation in the EU (adapted from data provided by the European Commission).

Seed	State of authorisation
Maize MON 810	Authorised – In process of renewal[1]
Maize 1507	In process
Maize Bt11	In process
Maize GA 21	In process
Maize 1507x59122	In process
Maize 59122	In process

[1] Maize MON 810 was approved by the European Commission on April 22, 1998 on the basis of the defunct Directive 90/220. This authorisation expired on April 18, 2007. It is currently considered to be an existing product, the 1998 authorisation being valid until the renewal application is resolved.

known as Amflora, after the MS did not manage to adopt a position via a qualified majority[27]. This authorisation was annulled by the ECJ at the request of Hungary, which was supported by other MS. Hungary alleged a manifest error in the precautionary principle which based the authorisation on deficient, incoherent and incomplete risk assessment (paragraph 66). The ECJ approved this claim, annulling the authorisation granted ruling that the Commission should have consulted the EFSA by virtue of the precautionary principle after being informed that the previous risk assessment report was incoherent (paragraph 103).

As well as these requests in process there were two GMOs whose authorisation for cultivation was requested and subsequently withdrawn by the company that owned them. This is the case of the Maizes BT11xMIR604xGA21 and MIR 604 whose authorisation applications were withdrawn on October 8, 2015. Syngenta, the applicant in both cases, based the decision to withdraw the applications presented in 2010 on criteria of profitability regarding them as economically unviable due to the fact that most MS have requested the exclusion of GMO cultivation from their territory (Swiss Info, 2015).

Given that the date of withdrawal of the authorisation application is subsequent to the transition period for notification of territorial restrictions, during this period there was the possibility of excluding the cultivation of 8 GMOs. The figure rose to 9 with the addition of the application for renewal of Maize MON 810. The MS enjoyed absolute discretion in deciding which GMOs they wished to exclude. However, there is a common pattern to be seen between the states that have requested some kind of exclusion.

With the exception of two states – Denmark and Luxembourg, whose case will be analysed below – all the MS that have restricted cultivation have done so for all GMO authorised or in the process of authorisation. These 17 states have notified the exclusion of the 8 possible crops, the approved MON 810, as well as the 6 in process at the end of the transition period. Moreover, all of them have also excluded the renewal application for authorisation of MON 810.

It is clear from this pattern that the intention of these MS – or of the regions in the case of Belgium and the UK – is not to exclude a specific GMO from cultivation, but to declare their territory free of GMO cultivation. The political component of such a decision is evident, so one would expect these MS to adopt a similar stance before future authorisation applications for cultivation. These states should not block the granting of new GMO authorisations, as the latter would not affect them because they could exclude them from their territory, thus permitting MS who so desired to proceed to their cultivation.

Denmark and Luxembourg have opted for more cautious activity. These states have only restricted the geographical scope of cultivation of Maizes MON 810 – as well as its renewal –, 1507, BT11 and GA21. In other words, they have not communicated restriction of Maize 1507x59122 or of 59122. In the case of Denmark, the reason for this decision is to be found in the modus operandi of the Danish

[27] The potato *Solanum tuberosum* L. line EH92-527-1 was approved via the Commission Decision 2010/135/EU of March 2, 2010.

Environment and Food Ministry. Denmark examines each application for GMO approval case by case, paying special attention to the EFSA scientific committee. The Danish government considers that it cannot take a definitive decision until the EFSA issues a conclusive opinion in order for it to make a decision. Therefore, they await this opinion before deciding whether or not to exclude their territory from cultivation. In accordance with article 26b of Directive 2015/412 (EU, 2015), Denmark will have 45 days following issue of this opinion in which to communicate a restriction

In the case of Maize 1507x59122 the EFSA has not yet published its definitive risk assessment report because the applicant was asked to supply new data. The process is suspended until the EFSA receives the information requested. With regard to maize 59122 the EFSA issued an additional report in November 2013 revising its previous report after the applicant communicated anomalies in the data. The EFSA communicated that it could not confirm the safety of the GMO until the applicant provided the missing information, and recommended that new analyses be carried out (EFSA, 2013). The doubts related to the effects on bees and coccinellidae.

Luxembourg has adopted the same modus operandi as Denmark and is waiting for the EFSA reports to decide on eventual territorial restriction of cultivation. Given its conduct in the past, it is highly probable that it will introduce restrictions once the EFSA issues its opinion on the GMOs in question.

3.5 Form of notification

Article 26b does not specify any formal requirement in order to request the adaptation of the geographical scope of the application for authorisation of cultivation. Therefore the form of the latter remains at the discretion of the MS. Though this is not made explicit, the understanding is that this will have to be in written form so the EC can fulfil its obligation to communicate this to the authorisation applicant, and to the rest of the MS. There is also an obligation to make this accessible to the general public via electronic means.

The MS have opted to issue individual notifications for each GMO authorised or in process of authorisation instead of issuing a joint notification, which was possible being a group of GMO of the same variety given that all the seeds are maize seeds[28]. As mentioned above, all the states that have requested adaptation of geographical scope have done so for Maize MON 810. As this seed is regarded as an existing product with a pending application for renewal, the MS states have communicated notifications vis-à-vis both circumstances. In this respect, there has been a difference in the way states have done this. Most have opted for a joint application embracing both MON 810 as an existing product and the renewal application. These states are Austria, Belgium, Bulgaria, Cyprus, Denmark, Germany, Latvia, Lithuania, Luxembourg, Malta and Slovenia. Meanwhile, Croatia, France, Greece, Hungary, Italy, the Netherlands and Poland have opted to issue separate notifications for the original authorisation and for the renewal application.

In the case of the UK the format has varied according to the territory affected. In the case of Scotland and Northern Ireland a joint notification was issued – per territory – for authorisation

[28] The notifications issued by Member States may be consulted online at: http://tinyurl.com/zxj9lcw.

and renewal. On the other hand, Wales opted for individual notification of restriction regarding original authorisation and the renewal application. This divergence in the form of notification reflects the existing territorial plurality. The UK has acted as a mere intermediary, sending a *Cover Letter* in which it announces that it issues the communication as requested by the administrations of these territories and submits the documents presented by them. Thus it is the Welsh administration that has decided to individualise notifications, unlike Scotland and Northern Ireland which have preferred to combine them as the same seed is under consideration.

Another aspect to bear in mind is the fact that many states have waited until the end of the 6-month period available to them in which to notify the exclusion of GMO cultivation from their territory. Of the 19 states that issued communications, only two – Latvia and Greece which submitted the notification in July – did so with more than two weeks' notice. The remaining states waited until the end of September to communicate their decision. The latest MS were Denmark, which communicated its restrictions two days before the deadline, and Malta and Luxembourg that did so with one day in hand.

The fact that states delay decision taking as much as possible is due to two factors. Firstly, many of them harboured doubts regarding the convenience, or not, of issuing the declaration owing to the political repercussions this might have. In many of these states there was national legislation that prohibited or hindered this cultivation, being these crops residual or non-existent. The other factor contributing to the delay is the intention of observing how other states act. These elements have contributed to the fact that most notifications have taken place within two weeks of the deadline. Avoiding the public repercussion of this communication is an understandable factor given that in the MS that have communicated restrictions there was no cultivation of these GMO. Therefore, the decision lacks practical significance given that currently no crops are affected. By contrast, the MS where they are cultivated let the deadline arrive without any notification whatsoever.

3.6 Justification of exclusions

The most striking aspect of the communication of geographical restriction of cultivation is that, with the notable exception of Germany, no MS justifies or offers any reason for such a decision. The MS have opted for a simple mention of the legal principle which enables them to adapt the territory of the authorisation and add that they communicate this in the appropriate manner. This lack of motivation seems to respond to the lack of will on the states' part to challenge the EFSA report, which has been favourable in every case, and to generate greater polemic with regard to an already controversial topic. Moreover, the MS and the Commission appear to have agreed not to questions the EFSA's scientific reasoning so as not to question the credibility of the institution. The system of approval and possible restriction of GMOs is designed to be as transparent as possible, with public dissemination of the application, the EFSA report, and the decision adopted by the European Commission. Furthermore, as has been stated previously, the territorial restrictions implemented by states are also communicated to the public. Therefore, the obfuscation of the MS regarding their reasons for adapting the geographical scope of GMO cultivation reduces the degree of transparency in the field and should be eradicated. The MS should not be afraid publicly to explain their reasons for adopting a particular policy.

This lack of transparency on the part of states when explaining the reasons behind their decisions may have consequences of another kind. So far all the notifications issued have been based on the first section of article 26b of the Directive. No member state has used the third section, which does require explanation of the motives chosen from amongst those that facilitate the norm and justification of these. The *modus operandi* of states seems to indicate that in the event of emergence of new applications for authorisation of GMO cultivation, states will invoke the procedure of the first section in order to avoid having to explain their decision. Today, it is difficult to imagine any MS resorting to Section 3 of article 26b. The most plausible option for this to occur is if there were to be a change of government in any MS that has not implemented any restriction, with a new government in favour of the latter. If no request has been made on the basis of Section 1, the only possibility is to resort to number 3.

To improve the levels of transparency in decision making, it would be desirable for the European Commission to sponsor a reform of the Directive, which will oblige an explanation of the request to adapt the geographical scope of an authorisation for cultivation.

As has been underlined, Germany is the only MS that has justified its decision. The German authorities argue that the cultivation of genetically modified maize is incompatible with traditional German agriculture, without further explanation of the meaning of this concept. For Germany the cultivation of genetically modified maize would increase the risk of natural maize being contaminated by the latter. This would necessitate the implementation of greater controls to detect any such contamination, which would lead to an increased workload and higher costs. In addition, they also based their decision on the protection of local biodiversity, as well as some particularly sensitive natural areas.

Germany also communicates that the restriction of cultivation will not be applicable to those holdings where it is carried out for purely scientific purposes, given that this would violate the freedom of scientific research included in article 5(3) of the Fundamental Law of Bonn, and protected by European legislation.

As we can see, the reasons given by Germany are not specific to a certain type of seed, but are applicable to all types of genetically modified maize, as well as to other possible seeds. Proof of this is that all the applications for exclusion presented by Germany contain the same reasoning, without any of them mentioning the seed in question.

Whilst the MS have not explained their decisions in their communications to the European Commission, some of them have done so via a public communication. This is the case of Slovenia, which announced its decision via a press release (MKGP, 2015). This confirmed that the decision falls within the directives of the agriculture and food plan that will be developed until 2020. The objective of the latter is to achieve sustainable development, agricultural products with greater added value, and preserve the environment and biodiversity. One of the measures to achieve these ends is the prohibition of GMO cultivation.

Although Hungary did not explain its restriction, its reasons are clear. Following the reform carried out in 2011, article XX of the Hungarian Constitution prohibits the cultivation of GMOs. The actual wording is as follows:

> ***Article XX***
>
> (1) Everyone shall have the right to physical and mental health.
>
> (2) Hungary shall promote the effective application of the right referred to in Paragraph (1) *by an agriculture free of genetically modified organisms*, by ensuring access to healthy food and drinking water, by organising safety at work and healthcare provision, by supporting sports and regular physical exercise, as well as by ensuring the protection of the environment. (Emphasis added).

The exception made by Hungary is but the consequence of the constitutional mandate of article XX. Although the new Constitution was unilaterally approved by the party in government, and strongly criticised by the opposition, there does appear to be consensus in the matter with broad political and popular support for prohibition.

The demand for restriction formulated by the Netherlands also contains particular elements. The communication to the Commission stated that the Dutch authorities are developing a political framework that will be used to decide upon the convenience, or not, of restricting the scope of application of authorisation of GMO cultivation. The government of the Netherlands has adopted these restrictions while this framework is being confirmed. The reason for this conditional restriction lies in the expiration of the transition period contemplated in article 26c. In the event of a decision to restrict cultivation once a decision has been made, it should invoke Section 3 of article 26b, which establishes a more laborious procedure, and requires a justification. Once the political framework is established, the authorities decide on each GMO case by case and, when appropriate, will maintain or eliminate the restriction.

In Denmark's case, the reason for exclusion is similar to that of the Netherlands. The Environment Committee of the Danish parliament decided to ask the government to implement the exclusions temporarily until parliament could carry out a thorough analysis of the question and its impact in Denmark (Folketinget, 2015). To this end it has requested technical reports evaluating the possible consequences of the Directive, in particular, in the event of coexistence of natural and genetically modified crops. Once these reports have been completed, the Parliament will decide whether to maintain or revoke the restriction.

Neither has Luxembourg explained its decision to restrict cultivation. It merely issued a press release referring to the reasons quoted in 2009 to activate the safeguard clause prohibiting the cultivation of maize MON 810. The reasoning offered by the Grand Duchy at that time was based upon protection of the environment, agriculture and the health of consumers. The same reasons were cited a year later to once again invoke the safeguard clause to prohibit cultivation of the Amflora potato (Santé Luxembourg, 2010).

3.7 Concluding remarks

On the basis of the above, one concludes that the changes introduced by Directive 2015/412 responded to an existing need, as there was a significant group of MS that sought to restrict the geographical scope of the applications for cultivation and required a legal framework in order to do so. The Directive has fulfilled this function, becoming a very important legislative instrument in the field of GMOs.

The Directive permits the coexistence within the European Union of zones where GMO cultivation is permitted along with zones where it is not. This disparity may raise doubts regarding the EU project as it has proved incapable of articulating a common solution. The states have divided into two groups: those that have introduced restrictions and those that have not. There have not been significant changes in the internal agricultural and environmental policy of the MS, as those which have requested exclusion from their territory are MS where these seeds are not cultivated. Meanwhile, the MS where these seeds were cultivated – as is the case of Spain, principal EU producer of transgenic maize – or which have an economic interest in them – the case of Ireland and feed for its cattle – are those that continue to permit cultivation in their territory.

Consequently, the legal framework introduced by Directive 2015/412 (EU, 2015) does not appear to have had an impact on the ground, as the MS are continuing with the same practices that they had been implementing prior to approval of the directive. The main objective of the law has been to endow existing situations with a legal architecture, clarifying the procedure to follow in order to avoid legal conflicts such as those resulting from the authorisation of the Amflora potato or the emergency measures introduced by France to restrict the cultivation of maize MON 810.

With regard to the GMOs excluded from cultivation, in this case there is greater homogeneity. Only Denmark and Luxembourg have not excluded cultivation of all possible GMOs. The rest have opted for a clear and firm policy of declaring their territory free of cultivation of these organisms. This pattern indicates a clear political purpose on the part of the MS who have requested adaptation of the geographical scope of the authorisations of cultivation. In the communications issued there is not – except in the cases of Germany and the Netherlands – any justification for taking this decision.

The lack of externalisation of the reasoning seems to indicate that these decisions are based on purely ideological grounds and not on empirical data. Despite the fact that the EFSA has published favourable opinions, none of these states have questioned these reports when issuing their notifications or have in any way questioned the institution's work. It would be desirable for MS to publish the reasons why they take the decision to restrict cultivation in their territory. If it were on the basis of scientific reasons other states might find these useful when making their own decision. They would also be of interest to the EFSA which should examine them with the goal of determining their impact and communicating this to all actors and the general public. If, on the other hand, the restrictions are based on ideological reasons with the objective of implementing certain public policies, the MS in question should communicate this publically for the sake of greater transparency facilitating greater accountability vis-à-vis their citizens.

Ideological reasons are equally valid in terms of compliance with the Directive, but it would be desirable for the MS to explain them. This is especially relevant in fields like food security, the environment and human and animal health where information must be clear and public to avoid alarmism and questions that lack any scientific basis.

The decision regarding GMO cultivation has been transferred to MS by Directive 2015/412 (EU, 2015) because the EU institutions consider this to be a sensitive question with special national, regional and local significance. MS have conveniently taken advantage of this opportunity evidencing a total absence of will when it comes to justifying their actions. The Commission should act to put an end to these practices in the interest of greater scientific-technical rigour and san improvement in levels of transparency and public governance.

The lack of justification is the main drawback of Directive 2015/412 (EU, 2015), which has on balance been positive. The success of this regulation has led the Commission to propose a replication of this model with regard to the use of GMOs for food and animal feed. In view of what has happened with cultivation the MS may very possibly opt for a variety of regulations so that diverse regulations are produced among MS. Prior to the introduction of this new legislative framework there should be analysis of the repercussions of the modifications introduced with regard to cultivation, and of the stances adopted by MS, with a view to contemplating the possible consequences of a framework of restriction of the use of GMOs.

Acknowledgements

Member of the Multilevel Constitutionalism Group, University of the Basque Country. Visiting researcher at the Centro de Estudios Políticos y Constitucionales, Madrid (Spain) and National University of Distant Education, Madrid (Spain).

References

Apolline, R. (2015). In the public interest? A comparative analysis of Norway and EU GMO Regulations. Reciel 24(3): 264-277.

Basque Government (BG) (2007). Council of the Basque Government Agreement of February 20, 2007.

Basque Government (BG) (2015). Non-binding Law 74/2015, on transgenics, plenary agreement, 10/11.02.01.0778. Official Bulletin of the Basque Government 132: 25200.

COGEM (2009). Socio-economic aspects of GMO. Building blocks for an EU sustainability assessment of GM crops. COGEM, Bilthoven, the Netherlands.

Christensen, R. (2015). Beretning afgivet af Miljøog Fødevareudvalget den 22. September 2015. Folketinget 2014-2015, Beretning 2.

ECJ (2013). General court ruling of December 13, 2013 related to issue T-240/10.

European Commission (EC) (2001). Directive 2001/18/EC of the European Parliament and of the Council of 12 March 2001 on the deliberate release into the environment of genetically modified organisms and repealing Council Directive 90/220/EEC. Official Journal of the European Union L 106: 1.

European Food Safety Authority (EFSA) (2010). Scientific Opinion on a question from the European Commission related to the notification by Portugal, pursuant to Article 95 (5) of the EC Treaty, for the prohibition of cultivation of Genetically Modified Plants in the Autonomous Region of Madeira. EFSA Journal 8(2): 1500.

European Food Safety Authority (EFSA) (2013). Scientific Opinion on an application from Pioneer Hi-Bred International and Dow AgroSciences LLC (EFSA-GMO-NL-2005-23) for placing on the market of genetically modified maize 59122 for food and feed uses, import, processing and cultivation under Regulation (EC) No 1829/2003. EFSA Journal 11(3): 3135.

European Union (EU) (2003a). Regulation (EC) no. 1.829/2003 of the European Parliament and of the Council of 22 September 2003 on genetically modified food and feed. Official Journal of the European Union L 268.

European Union (EU) (2003b). Regulation (EC) no. 1.830/2003 of the European Parliament and of the Council of 22 September 2003 concerning the traceability and labelling of genetically modified organisms and the traceability of food and feed products produced from genetically modified organisms and amending Directive 2001/18/EC. Official Journal of the European Union L 268.

European Union (EU) (1990). Council Directive 90/220 of 23 April 1990 on the deliberate release into the environment of genetically modified organisms (90/220/EEC). EU, Brussels, Belgium.

European Union (EU) (2015). Directive (EU) 2015/412 of the European Parliament and of the Council of 11 March 2015 amending Directive 2001/18/EC as regards the possibility for the Member States to restrict or prohibit the cultivation of genetically modified organisms (GMOs) in their territory. Official Journal of the European Union L 68: 1.

Fischer, K., Ekener-Petersen, E., Rydhmer, L. and Björnberg, K.E. (2015). Social impacts of GM crops in agriculture: a systematic literature review. Sustainability 7(7): 8598-8620.

Flemish Government (2010). Executive order of the Flemish Government establishing general measures for the coexistence of genetically modified crops and conventional and organic crops (Oct. 15, 2010), art. 3, M.B., Nov. 30, 2010, 73,420 and Flemish Executive Order of Oct. 15, 2010, for the Coexistence of Crops, art. 4.

Government of Northern Ireland (GNI) (1998). Northern Ireland Act 1998, c.47. Government of Northern Ireland, Dublin, UK.

Government of Scotland (GS) (1998). Scotland Act 1998, c.46 and its respective modifications by the Constitutional Reform Act 2005, c.4 and the Scotland Act 2012, c.11. Government of Scotland, Edinburgh, UK.

Government of Wales (GW) (2006). Government of Wales Act, c.38 modified by the Government of Wales Act 2006, c.32. Government of Wales, Cardiff, UK.

Greiter, A., Miklau, M., Heissenberger, A. and Gaugitsch, H. (2011). Socio-economic aspects in the assessment of GMOS – Options for action. Environment Agency (Umweltbundesamt) Austria, Vienna, Austria, 52 pp.

Republic of Slovenia, Ministry of Agriculture, Forestry and Food (MKGP) (2015). Slovenia bans cultivation of maize varieties MON810 (approved and in re-authorisation procedure), 1507, Bt11, GA21, 59122, MIR604, Bt11xMIR604xGA21 and 1507x59122. Available at: http://tinyurl.com/jzchhem.

Roger, A. (2015). In the public interest? A comparative analysis of Norway and EU GMO regulations. Reciel – Comparative Environmental Law 24(3): 264-277.

Santé Luxembourg (2010). Pomme de terre OGM: le Ministre de la Santé fait usage de la clause de sauvegarde. Available at: http://tinyurl.com/j6p7q3y.

Spök, A. (2010). Assessing socio-economic impacts of GMOs. Issues to consider for policy development. Bundesministerium für Gesundheit, Vienna, Austria.

Swiss Info (2015): Syngenta Scraps European Applications for GMO Maize Seeds. 14 October, 2015. Available at: http://tinyurl.com/yc2tnxch.

4. Labelling requirements of GM food as compared to those of other foods

A. Arroyo Aparicio
Professor of Commercial Law, Faculty of Law, National University of Distant Education (UNED), c/Obispo Trejo 2, 28040 Madrid, Spain; aarroyo@der.uned.es

Abstract

The title of this paper alludes to two aspects relating to the purpose of consumer information requirements for foodstuffs in general and the rules specifically applying in the case of genetically modified products. This is an extraordinarily broad subject, in which we shall attempt her to focus on questions of private law, and an intense debate has been sparked in the EU with regard to what information consumers should have, for the purposes of food safety, but more especially, to allow them to make an informed choice. The various approaches concern consumers' doubts about this type of food, their possible rejection of it and the idea that it might constitute 'unnatural' food, although this is not a legal debate, in the strict sense – or certainly not a uniquely legal one. Another interesting issue is notion of what constitutes a consumer; however, we shall delve into this discussion in any great depth here, basing our discussion on the general – abstract – notion of the consumer contained in general regulations as a whole. There are also specific types of consumer – the final consumer in the case of the retail trade, for example, with reference to certain specific acts and the average consumer, in the case of the consumer who receives information through a variety of means (advertising or labelling). In the context of private law, it is certainly considered that the specific consumer can exercise actions on the basis of the regulation considering unfair practises towards consumers to be unlawful; and associations of consumers may take collective action in the case of damage to a given group of people.

Keywords: labelling of GMO food, unfair commercial practices, consumer food information, informed choice

4.1 Background and purpose of consumer food information rules

The purpose of any food labelling regulation is that consumers should have essential, legible and understandable information, using which they can make an informed choice on the purchase of the products in question. This apparently straightforward notion is in contrast with the extensive and complex regulation on the issue; for it is important to bear in mind that in the matter of foodstuffs the very complexity of the product itself also entails additional requirements.

The basic general regulation on consumer food information is Regulation (EU) 1169/2011 (EU, 2011) of the European Parliament and of the Council, of 25 October 2011 on the provision of food information to consumers (Borghi, 2015; Bremmers, 2012; Costato, 2013; Nihouil and Van Nieuwenhuyze, 2012). Two aspects relating to the scope of this regulation need to be borne in mind. Firstly, the Regulation applies to food business operators at all stages of the food chain and to all foods intended for the final consumer, including foods delivered by mass caterers, and foods intended

DOI 10.3920/978-90-8686-845-2_4, © Wageningen Academic Publishers 2017

for supply to mass caterers. And secondly, the Regulation applies without prejudice to labelling requirements provided for in specific Union provisions applicable to particular foods.

With regard to the general requirements of this Regulation, its aim is that *neither* the labelling, *nor* the presentation of the food products, *nor* any advertising regarding same may:

- mislead the consumer as to the characteristics, properties or effects;
- attribute to any food the property of preventing, treating or curing a human disease, (with the exception of natural mineral waters and foodstuffs intended for particular nutritional uses, for which specific provisions exist).

These aims will be achieved if the food information is accurate, clear and easy to understand for the consumer.

Having established this primary purpose, if one looks beyond the most immediate protection – labelling and foodstuffs – to the area of general protection, one must start from the idea of the protection of the consumer on the internal market (on general policy, see Guillén Caramés, 2002). Within that general framework, it should be remembered that a consumer who acquires a product in any State of the EU will expect it to meet certain common requirements. Indeed, the aforementioned regulation takes as its legislative base Article 169 of the Treaty on the Functioning of the European Union which states that the Union is to contribute to the attainment of a high level of consumer protection by the measures it adopts pursuant to Article 114. In other words, as also stated in Regulation 1169/2011, the free movement of safe and wholesome food is an essential aspect of the internal market and contributes significantly to the health and well-being of citizens, and to their social and economic interests. Within the regulatory area of the EU, in order to achieve a high level of health protection for consumers and to guarantee their right to information, it should be ensured that consumers are appropriately informed as regards the food they consume. It is also true, that consumers' choices can be influenced by, *inter alia*, health, economic, environmental, social and ethical considerations.

In the matter of food, for several decades, the EU has gradually adopted a series of rules pertaining not only to food labelling, but also on permitted additives, safety and hygiene and, of course, on control.

In this context, it is important to take into account Regulation (EC) No 178/2002 of the European Parliament and of the Council of 28 January 2002 (EU, 2002), laying down the general principles and requirements of food law, establishing the European Food Safety Authority (dealt with elsewhere in this text) and laying down procedures in matters of food safety. As stated in Regulation 178/2002, a general principle of food legislation is to provide consumers a basis to make informed choices in relation to the foods they consume and prevent any practices that may mislead the consumer.

This 'informed choice' by consumers, that is, the minimum information they need to know is specified in the matter of labelling with reference to the mandatory particulars that operators must include under Regulation 1169/2011 (essentially: the name of the food, the list of ingredients, the net quantity, the 'use by' date, any instructions for use that are necessary, the name and address of the operator and the nutritional information). This Regulation, which came into force in 2014, has been amended by others subsequent texts (applicable from 13 December 2014, with the exception of the provisions on nutritional information, applicable from 13 December 2016).

Questions that aroused particular debate when this Regulation was enacted were those relating to: (1) the amount of 'added water' with regard to which the name of the food must include an indication of its presence, if the added water makes up more than 5% of the weight of the finished product; (2) engineered nano materials ('nano' information that must be targeted at the 'average consumer' and not at a vulnerable group or group worthy of specific consumer protection (Bremmers, 2012; Leible, 2010); and (3) the labelling referring to the origin or place of provenance.

Under Regulation 1169/2011, the mandatory particulars must be made available to consumers using distance selling to purchase food before they conclude their purchase at no additional cost. Thus, in the case of products that are offered for sale by means of distance communication (Cf. Articles 8 and 14) – a clear manifestation of the internal market in consumer matters (Arroyo, 2003) – the food business operator responsible for the food information is the operator under whose name or trading name the food is marketed.

As for the amendments, it is worth noting in particular that introduced by the regulation on 'novel foods', Regulation (EU) 2015/2283 (EU, 2015) of the European Parliament and of the Council of 25 November 2015, amending Regulation (EU) No 1169/2011 and repealing Regulation (EC) No 258/1997 (EU, 1997) and Regulation (EC) No 1852/2001(EU, 2001). The regulation came into force on 1 January, 2016 and will mostly be applicable from 1 January, 2018, when Regulation (EC) 258/1997 and the Regulation (EC) 1852/2001 will be repealed.

The Regulation on 'novel food' (in general, see González Vaqué, 2015) determines the kinds of food classified as novel. Essentially, this comprises: food not consumed in the EU in significant amounts before May, 1997, when the first regulation on 'novel food' came into force, which may consist of innovative foods that have been developed or produced with new technologies and procedures, and foodstuffs that are traditionally eaten outside the EU. However, genetically modified foods *already regulated* by others acts of the Union do not fall within the scope of application, hence genetically modified foods included within the scope of Regulation (EC) No 1829/2003 (EU, 2003a) of the European Parliament and of the Council are not subject to the 2015 Regulation.

It is worth noting that, in general, 'novel foods' are subject to the general labelling requirements established in Regulation 1169/2011. Likewise, specific additional labelling requirements may be applied to inform consumers adequately, as is the case of labels warning that a novel food is not recommended for certain vulnerable groups. The label must contain the name of the food and, where necessary, the conditions of use. Any nutritional and health claims must also comply with Regulation (EC) No 1924/2006 (EU, 2006).

Finally, it should be said that when it was enacted, Regulation 1169/2011 was widely considered to be a *missed* opportunity to have regulated in greater depth the labelling of genetically modified products (Bremmers, 2012) which continued to be specifically subject to the sectorial regulation (*GMO regulation package*), in particular, to Regulation 1829/2003, in combination with the rules on '*the deliberate release into the environment of genetically modified organisms*' (Directive 2001/18).

4.2 Labelling of genetically modified products

As already mentioned, the present Regulation on 'novel food' does not apply to genetically modified products falling within the scope of Regulation 1829/2003 (Art. 2.2.a, V in the 2015 Regulation).

However, genetically modified foods are not always deemed to lie outside the classification of 'novel food'. Indeed, the former Regulation (EC) 258/1997 – which although repealed remains in force until 31 December 2017 – specified what should be classified as 'novel food' (foods that were not in widespread use before its entry into force) and provided for a safety assessment before they were marked within the Union (for a general discussion of this Regulation see González Vaqué, 1997). There is a clear difficulty in comparing the former 1997 Regulation with the 2015 Regulation on novel food, among other reasons because the determination of what comprises novel food varies considerably. The definition in the old Regulation was much more generic; the difference is probably not only due to technological advances, but also because there has since been a move towards a more precise definition.

In particular, Article 8 of Regulation 258/1997 was dedicated to consumer information and contained a statement that sparked an intense debate with regard to the labelling of transgenic foods. The Regulation stated that the labelling must give information on any characteristic which renders a novel food or food ingredient no longer 'equivalent to an existing food or food ingredient'. Specifically, a clause was included whereby, for the purposes of the article a novel food or food ingredient was deemed to be no longer equivalent if 'scientific assessment, based upon an appropriate analysis of existing data' could demonstrate that 'the characteristics assessed' were different in comparison with a conventional food or food ingredient, having regard to the accepted limits of natural variations for such characteristics.

The system designed under the 1997 Regulation resulted in disparities between the relative levels of requirement of the different Member States. This led to the enactment of the 2015 Regulation on 'novel food', drafting of which began in 2013. The recitals of the new Regulation state that 'Differences between national laws concerning the safety assessment and authorisation of novel foods may hinder the free movement of such food, thereby creating legal uncertainty and unfair conditions of competition'. As discussed, in all cases, food intended to be used for technological purposes and genetically modified food which is already covered by other Union acts is excluded from the scope of this Regulation. Therefore, genetically modified foods that *do* fall within the scope of Regulation (EC) No 1829/2003 of the European Parliament and of the Council are governed by it and not by the 2015 Regulation.

Thus, in essence, in addressing the labelling of genetically modified foods, the most immediate references area Regulations 1829/2003 and 1830/2003 (EU, 2003b). The regulations cover the labelling, monitoring of risks for the environment and health, and the capacity, where necessary, to withdraw the products from the market. The first Regulation regulates how GMOs are to be authorised and supervised and how genetically modified foodstuffs and animal feed are to be labelled. The second Regulation, 1830/2003, establishes the rules to ensure traceability of products that contain GMOs and food and feed products obtained from GMOs throughout all stages of the production and distribution chains.

Traceability (the ability to track GMOs and products produced from GMOs at all stages of the production and distribution chain) is stated to be key in providing consumers with information. Three principal requirements are established for sellers: (1) to inform trade buyers in writing that a product contains GMOs (or provide a 'declaration of use' for products intended for food or animal feed); (2) to communicate the unique identifiers assigned to each GMO under the regulation; and (3) (for food and feed) to identify each ingredient produced from GMOs, if an ingredients list exists. This information must be provided at every stage in the production and distribution chain and kept for five years.

As for labelling, in general, final consumer packaging or pre-packaged products containing GMOs should be labelled: '*This product contains genetically modified organisms (or the names of the organisms)*' However, exemptions do exist. These are as follows: products that contain traces of GMOs (below 0.9%) if this is technically unavoidable (and medicinal products).

Three general aspects merit highlighting. Firstly, these regulations are based on a definition of genetically modified food and some techniques of genetic modification lie outside the scope of the regulation – although where they are classed as 'novel food', they are covered by the 2015 Regulation, and by the general regulations on labelling (Chapter 4.2).

Secondly, all ingredients (including additives and flavouring) which are, contain or are produced from GMOs, are subject to the rules on labelling (irrespective of whether or not they contain DNA or proteins derived from genetic modification, since in the new Regulation 1829/2003, labelling is not based on the detection of GM DNA or proteins). Therefore, the products derived from GMOs, even if they have undergone a process of transformation that eliminates the genetically modified DNA or protein, are also subject to the rules on labelling.

And thirdly, organic foods are subject to specific regulation whereby all forms of GMOs are prohibited. In other words, under Regulation 834/2007 (EU, 2007), given that GMOs and products produced from or by GMOs are incompatible with the concept of organic production and consumers' perception of organic products, they should not be used in organic farming or in the processing of organic products.

In the matter of GMOs, one possible option is to label products as being GM-free. In 2013, this voluntary option was the subject of an external study – drawing on extensive data – which is available on the European Union website. The report found that there was a lack of uniformity with regard to the information provided, the type of label and the impact on the consumer's choice.

Another issue that needs to be considered is the difference that has always existed in the treatment of transgenic products between the American and European models. In broad terms, the difference has been defined has being between an approach centring on the final result or product (in the US) and an approach based on the process and control thereof (EU) (Bruton, 2013).

In Europe, as has been discussed, the aim is that any product that contains or consists of GMOs, and any food produced from GMOs is subject to authorisation, supervision and labelling applicable by virtue of Regulation 1829/2003, with certain specified exemptions (non-intentional presence not exceeding a threshold of 0.9%). The exemptions in themselves and their practical determination may certainly be subject to debate, as has been seen in some Court of Justice of the European Union (CJEU) cases (case 442/09, of 6 September 2011). In other cases the CJEU has addressed more specifically the possible misleading of consumers (sentence in case C-195/14 regarding labelling of foodstuffs and the misleading of consumers as to the composition thereof with regard to Directive 2000/13/EC of 20 March 2000 (EU, 2000), relating to the labelling, presentation and advertising of foodstuffs).

In its 2011 judgement, the CJEU stated that a substance such as the pollen derived from a genetically modified variety of corn that has lost its ability to replicate and which lacks any capacity to transfer the genetic material it can no longer be categorised as a GMO. However, products such as honey and food supplements containing such a substance constitute foodstuffs containing ingredients produced from GMOs and that classification may be made irrespective of whether the substance was incorporated intentionally or adventitiously. In the judgement of the Court, this means that there is a requirement for supervision and authorisation of said products by the food authority. This case showed that the coexistence of traditional and transgenic crops is a complicated issue and may require the enforcement of costly checks in certain circumstances (such as that of honey producers).

In the US, on the other hand, lawsuits have been taken in recent years against economic operators over the advertising of products containing GMOs as being 'all natural'. In this context, it has been seen that even where the Food and Drug Administration has not recognised the requirement for specific labelling, the courts in many such cases ruled that foodstuffs with GMOs 'are inherently unnatural' (Negowetti, 2014).

4.3 Relationship with regulations on unfair commercial practices

Regulation 1169/2011, Recital 5 makes reference to Directive 2005/29/EC of the European Parliament and of the Council, of 11 May, 2005 (EU, 2005), concerning unfair business-to-consumer commercial practices in the internal market. This Directive covers certain aspects of the provision of information to consumers specifically to prevent misleading actions and omissions of information. Recital 5 states that the '*general principles on unfair commercial practices should be complemented by specific rules concerning the provision of food information to consumers*'.

In accordance with this idea, consumer protection regulations on this matter – unfair commercial practices – also broadly protect consumers of foodstuffs, without prejudice to specific rules on food information. Thus they protect the economic interests of consumers before, during and after the completion of a commercial transaction.

Unfair commercial practices are those which: do not comply with the requirements of professional diligence; are likely to materially distort the economic behaviour of the average consumer. Moreover, certain population groups must be particularly protected due to their particular vulnerability to the practice or the product, because of their age (children or the elderly), credulity or mental or physical infirmity. The Directive defines two specific categories of unfair commercial practice: misleading practices (by action or omission) and aggressive practices.

As for application of the Directive, it sets out a catalogue of prohibitions (or 'black list'), which should constitute the starting point for examining a case. Under the system adopted, the means of determining whether or not a given commercial practice is prohibited by the Directive is as follows. Firstly, it must be determined whether it falls within any of the practices on the 'black list' contained in Annex I of the Directive; if not, it must be determined whether it constitutes a misleading (Articles 6 and 7 of the Directive) or aggressive (Articles 8 and 9) practise; and if not, it must be determined whether it infringes the general clause of Article 5 of the Directive. The Directive was incorporated into Spanish law in the Unfair Competition Act (*Ley de Competencia Desleal*) (Arroyo, 2011), i.e. as situations contrary to the corrective actions on the market and prosecutable by the ordinary courts to obtain, *inter alia*, compensation for the damages suffered.

Of the unfair practices that listed in the Directive, there are several that might apply in this matter: (1) the displaying of a trust mark, quality mark or equivalent without having obtained the necessary authorisation. (Appendix I, 2); (2) claiming that a product has been approved, endorsed or authorised by a public or private body when it has not or making such a claim without complying with the terms of the approval, endorsement or authorisation (Appendix I, 4); (3) stating or otherwise creating the impression that a product can legally be sold when it cannot. (Appendix I, 9); (4) misleading actions regarding the geographical origin (Art. 6); (5) falsely claiming that a product is able to cure illnesses, dysfunction or malformations.

With regard to this last issue, mention should be made of Regulation 1169/2011, which refers to this issue in similar terms. Article 7.3 of the Regulation establishes that subject to derogations provided for by Union law applicable to natural mineral waters and foods for particular nutritional uses, food information shall not attribute to any food the property of preventing, treating or curing a human disease, nor refer to such properties. As already mentioned, this regulation applies to all operators throughout the food chain.

This issue may also be related to the regulation on nutrition and health claims made on foods. This Regulation sets out the rules on declarations and properties in commercial communications such as advertising. Declarations may not be made without prior authorisation.

As can be seen, the connection between GM food labelling and unfair commercial practices is not entirely direct, although this Directive, its national transposition rules and the procedural means related to the two may certainly constitute one more means of ensuring that consumers obtain information that is correct and not misleading.

4.4 Final reflections

The legislation on food information in general and the specific legislation referring to transgenic food must take into consideration two consumer interests: the right to information and the freedom of choice. The former is considered essential for the exercise of the latter; if the information is not adequate, it is difficult for any 'informed choice' to be made.

In the case of transgenic foodstuffs, there are reasons other than those applying to other foodstuffs for requiring specific information. The development or evolution of this matter is also related to the technical complexity of the processes of genetic modification with which the average consumer is certainly not familiar. For this reason the labelling of food as GMO-free might be of help, since it would give consumers the chance to opt-out, rejecting foods that do not display such labels. As discussed, such labelling is voluntary and, according to the comparative study conducted, the degree of implementation varies considerably amongst Member States.

With some differences, this *ex facto* situation (different levels of protection) also occurred *ex lege* when minimum harmonisation directives in the EU on the protection of consumers' economic interests first began to be implemented. For this reason, it was established at the outset of minimum harmonisation that States with more protective – or advanced – legislations could go further. At present this legislative option has gradually been replaced by that of full harmonisation directives, in that the harmonised standard reached is, in general, now greater (CF. the Directive on unfair practices, most of whose provisions are full harmonisation clauses).

References

Arroyo Aparicio, A. (2003). Los contratos a distancia en la Ley de Ordenación del Comercio Minorista. Aranzadi, Pamplona, Spain.

Arroyo Aparicio, A. (2011). El impacto de la Directiva sobre Derechos de los Consumidores en el ordenamiento español, Actualidad jurídica Aranzadi, 867, 2013, p 16; with the updating of Spanish legislation: 'Contratos a distancia y fuera del establecimiento mercantil, celebrados con consumidores, según la Ley General para la Defensa de Consumidores y Usuarios, Contratos Mercantiles (A. Bercovitz), published in 2017; 'Prácticas Agresivas', Comentarios a la Ley de Competencia Desleal, A. Bercovitz (Dir.), Aranzadi, Pamplona, Spain, pp. 197-221.

Borghi, P. (2015). El nuevo marco normativo europeo relativo al etiquetado y la información al consumidor. Lecciones de Derecho alimentario 2015-2016. Thomson Reuters Aranzadi, Pamplona, Spain, pp. 89-109.

Bremmers, H. (2012). An integrated analysis of food information to consumers: problems, pitfalls, policies and progress. In: Rickert, U. and Schiefer, G. (eds.) Proceedings in food system dynamics. University of Bonn, Bonn, Germany, pp. 614-627.

Bruton, S. (2013). GMO food labelling. Encyclopedia of Food and Agricultural Ethics, Springer, New York, NY, USA.

Costato, L., Borghi, P. and Rizzioli, S. (2013). Compendio di diritto alimentare. CEDAM, Wolters Kluwer, Alphen aan de Rijn, the Netherlands, pp. 220-233.

European Union (1997). Regulation (EC) No 258/97 of the European Parliament and of the Council of 27 January 1997 concerning novel foods and novel food. Available at: http://tinyurl.com/y7tec7l5.

European Union (2000). Directive 2001/13/EC of the European Parliament and of the Council of 26 February 2001 amending Council Directive 95/18/EC on the licensing of railway undertakings. Available at: http://tinyurl.com/yc5crcvs.

European Union (2001). Commission Regulation (EC) No 1852/2001 of 20 September 2001 laying down detailed rules for making certain information available to the public and for the protection of information submitted pursuant to European Parliament and Council Regulation (EC) No 258/97. Available at: http://tinyurl.com/yarbfmhp.

European Union (2002). Regulation (EC) No 178/2002 of the European Parliament and of the Council of 28. January 2002 laying down the general principles and requirements of food. Available at: http://tinyurl.com/np2pbzm.

European Union (2003a). Regulation (EC) No 1829/2003, genetically modified food and feed. Available at: http://tinyurl.com/y7b255p8.

European Union (2003b). Regulation (EC) No 1830/2003 concerning the traceability and labelling of genetically modified organisms and the traceability of food and feed products. Available at: http://tinyurl.com/y94hrdwp.

European Union (2005). Directive 2005/29/EC of the European Parliament and of the Council of May 2005 Concerning unfair business-to-consumer commercial practices in the internal market. Available at: http://tinyurl.com/cgfxbkl.

European Union (2006). Regulation (EC) no 1924/2006 of the European parliament and of the council of 20 December 2006 on nutrition and health claims made on foods. Available at: http://tinyurl.com/dfkbbo.

European Union (2007). Council Regulation (EC) No 834/2007 of 28 June 2007 on organic production and labelling of organic products. Available at: http://tinyurl.com/4dcvy8.

European Union (2011). Regulation (EC) No 1169/2011 on the provision of food information to consumers. Available at: http://tinyurl.com/ya7d4xsy.

European Union (2015). Regulation (EU) 2015/2283 of the European Parliament and of the Council of 25 November 2015 on novel foods. Available at: http://tinyurl.com/y96zj4ta.

González Vaqué, L. (1997). El Reglamento (CE) No. 9 258/97 sobre nuevos alimentos: información del consumidor y evaluación de su seguridad. European Commission, Brussels, Belgium.

González Vaqué, L. (2015). Qué hay de 'nuevo en el Reglamento (UE) 2015/2283 relativo a los nuevos alimentos? Revista de Derecho Agrario y Alimentario 68: 107.

Guillén Caramés, J. (2002). El estatuto jurídico del consumidor. Civitas, Madrid, Spain, pp. 71.

Leible, S. (2010). Consumer information beyond food law. European Food and Feed Law Review 5(6): 316-324.

Negowetti, N.E. (2014). Food labeling litigation: exposing gaps in the FDA's resources and regulatory authority. Brookings Institution, Washington, DC, USA.

Nihoul, P. and Van Nieuwenhuyze, E. (2012). L'étiquetage des denrées alimentaires: une pondération réussie entre intérêts contradictoires? Journal de Droit Européen 20(192): 237-243.

5. A scientific view of the current status of genetically modified crops, foods and feed in the European Union

M. De Renobales
Professor of Biochemistry and Molecular Biology, Faculty of Pharmacy, University of the Basque Country, Paseo de la Universidad, 01006 Vitoria-Gasteiz, Spain; mertxe.derenobales@ehu.eus

Abstract

The present chapter examines the current situation of genetically modified (GM) crops in Europe from a scientific point of view. First of all, it briefly presents techniques to introduce desired characteristics into crops, or remove adverse ones, including conventional, genetic engineering and gene-editing techniques. The definition of GM plants given in Directive 2001/18, although legal, is not a scientific definition because all techniques used to improve crop plants introduce heritable genetic modifications. That definition does not apply to plants improved by the recent gene-editing techniques in the European Union. An analysis of the presently commercialised GM plants in the EU indicates there is a *de facto* blockage of authorisation for cultivating GM plants. The difference in timing to approve GM crops between the EU and other countries from which GM varieties can be imported ('asynchronous approval') is causing serious difficulties to the still strong livestock sector, and it threatens to increase malnutrition in the economically more vulnerable population sector. A succinct analysis of the factors leading to the low level presence of unapproved GM crops indicates that these are easily singled out because transgenic events are easily detectable, even in trace amounts. This ease of detection is at the heart of the social debate and regulatory procedures surrounding the GM technology, including coexistence with non-GM crops. Finally, a brief analysis of the contributions of GM crops to the EU consumer and the environment is offered.

Keywords: plant enhancing techniques, genetically enhanced plants, commercialised GM plants in the EU, asynchronous approval, coexistence

5.1 Introduction – Plant enhancing techniques and plant genome modifications

Genetic improvement, or enhancement, of plants consists of obtaining a plant with a useful characteristic(s) that it did not have before and which can be passed on to its descendants. For that to occur, that characteristic must be encoded in the plant's genome. Some examples of characteristics useful for various sectors of the population are the following:

- for farmers:
  - resistance to biotic stresses: pests of any kind (herbivores, virus, bacteria, nematodes, etc.), including tolerance to herbicide to fight weeds which use up water and soil nutrients;
  - resistance to abiotic stresses: drought; saline, alkaline and acidic soils; extreme temperatures;
  - traits to increase plants' potential yields: improved use of resources (fertilisers), improved photosynthesis.
- for consumers:
  - improved nutrient content: macro- and micro-nutrient levels, healthier fatty acid oil profiles;

DOI 10.3920/978-90-8686-845-2_5, © Wageningen Academic Publishers 2017

    - reduced toxin levels, both endogenous and those produced during cooking or processing;
    - improved sensory properties.
- for processors:
    - delayed ripening, higher concentration of solids (less thickening agents added).

To achieve those (and other) improvements a variety of techniques have been, and are, currently used (Cubero, 2013). The general public, as well as many scientists not involved in plant genetic improvement, is primarily familiar with the technique of hybridisation (or crossing) between sexually compatible species. Other frequently used techniques are forced hybridisation between sexually incompatible species followed by embryo rescue, tissue culture techniques, modification of the number of chromosomes (polyploidy), artificially induced mutagenesis by chemical compounds and/or irradiation, genetic engineering and, recently, genome editing.

All these techniques, except for the mutations that occur spontaneously in nature, are realised in controlled environments such as a greenhouse and the laboratory.

Radiation-induced mutagenesis is a remarkable technique, perhaps the most surprising one, developed by the FAO and the International Atomic Energy Agency (IAEA) in the 1950s to accelerate the rate of appearance of mutations in plants. The objective was '*to contribute...to peace, health and prosperity throughout the world, especially to global food security and sustainable agricultural development*' (FAO-IAEA, 2016a). Plants or seeds are irradiated to accelerate the appearance of mutations in their genes. Plants are grown for several generations and those with the desired characteristic(s) are selected (FAO-IAEA, 2011). In the FAO-IAEA database, as of late May 2016, there were 3,234 varieties of the most frequently used edible plant species such as rice, barley, wheat, maize, various other cereals, legumes, vegetables, oil crops, as well as industrial crops and ornamentals (FAO-IAEA, 2016b). In the last 7 or 8 months 16 newly developed varieties have been added to the database, clearly indicating that today it is an actively used technique.

During the 70s and 80s of the past century, genetic engineering techniques allowed scientists and plant breeders to introduce a foreign gene, or genes, from widely distant plant or organism into the target plant (Slater *et al.,* 2008). These techniques permit scientists to use the most suitable combination of the three regions of a gene (promoter, coding zone and terminator) to achieve the intended level of expression of the inserted gene in the improved plant. Often, these three regions come from different organisms and are assembled together in the laboratory and inserted into the target plant (transgenesis). In some cases, the three elements (promoter, gene with its introns, and terminator) are isolated as a whole from a sexually compatible plant and inserted in the same order in which they occur without any modification (cisgenesis). Finally, these three elements from a sexually compatible plant can be rearranged and/or modified before being inserted into the target plant (intragenesis).

Recently, a new group of techniques, collectively referred to as 'genome editing' techniques, have been developed that permit the introduction of very specific modifications into the plant's genome, sometimes modifying only a few bases (Doudna and Charpentier, 2014). Quite often the distinction between a plant improved by these techniques and a plant improved, for example, by induced mutagenesis may not be possible. These techniques are currently being used to improve a variety of crop plants.

All these techniques, without exception, introduce genetic modifications in the plants' genomes, adding beneficial traits or removing detrimental ones. All these changes can be transmitted to the next generation. Induced mutagenesis by treatment with chemical mutagens or ionising radiation is the method that causes largest genetic disruption in the genome of the treated plant. Likewise, the likelihood of introducing negative unintended effects, although variable, is also largest in the case of induced mutagenesis (National Research Council, 2004).

5.2 The EU and genetically enhanced plants

From a legal and regulatory point of view, the EU considers that the possibility of a given crop plant to harm consumers and/or the environment depends exclusively on the technique(s) used to enhance the characteristics of that crop, with genetic engineering techniques being the only ones capable of causing problems. Current legislation reflects this thinking even though it is not stated anywhere, to the best of my knowledge. Food plants are, thus, divided into two unequal groups depending on the techniques and processes used to improve their agronomic, nutritional and/or processing characteristics: (1) the few plants improved by the use of genetic engineering techniques, and (2) all the rest of crop plants improved by the use of the rest of the techniques mentioned in the previous section of this chapter.

According to Art. 2 of Directive 2001/18 (European Parliament, 2001), a 'genetically modified' (GM) organism is, legally, '*an organism, with the exception of human beings, in which the genetic material has been altered in a way that does not occur naturally by mating and/or natural recombination*' (art. 2). Annex 1A part 1 lists all techniques which result in GM plants from a legal and regulatory point of view:

1. recombinant nucleic acid techniques involving the formation of new combinations of genetic material by the insertion of nucleic acid molecules produced by whatever means outside an organism, into any virus, bacterial plasmid or other vector system and their incorporation into a host organism in which they do not naturally occur but in which they are capable of continued propagation;

2. techniques involving the direct introduction into an organism of heritable material prepared outside the organism including micro-injection, macro-injection and micro-encapsulation;

3. cell fusion (including protoplast fusion) or hybridisation techniques where live cells with new combinations of heritable genetic material are formed through the fusion of two or more cells by means of methods that do not occur naturally.

In this article I will refer exclusively to plants which can be used as food *per se,* and/or to provide raw materials or ingredients for food and feed.

This definition of a GM plant strongly implies that only those techniques previously listed (Annex 1A part 1) modify plant genomes in a manner that the modifications introduced can be continually propagated, but the other techniques do not. From a legal point of view, the group of plants improved by the so-called 'conventional' techniques are not 'genetically modified' organisms, implying that their genomes have not been modified, and thus inducing the general public to think that plants improved by those techniques (Annex IA part 2 and Annex III) do not carry any genetic modifications and are, therefore, 'natural'. From a scientific point of view, this distinction based on techniques and processes to improve the characteristics of the plant is arbitrary and not science-based, because all crop plants that feed us have been genetically modified, most of them throughout centuries or millennia, as compared to their wild relatives. For the consumer and the environment, it is the final improved plant with its specific phenotype what matters, because the improved plant is the one that will be grown in the open field, processed into foods and feed, and ultimately eaten. But the legal distinction between GM and conventional (non-GM) has very important regulatory, social and economic implications.

Conventional crop improving techniques are certainly capable of introducing harmful characteristics. A survey of the published literature provides us with, at least, four examples of crops improved by crossing with sexually compatible wild relatives to improve the resistance of the cultivated plant to harmful insects by hybridising it with resistant wild plants. No other techniques were used. The final plants were discovered to be toxic after their being commercialised due to excessively high levels of naturally occurring toxic compounds[29]: the Lenape potato (Akeley *et al.,* 1968), a celery (Seligman *et al.,* 1987) and two squashes (Rymal *et al.,* 1984). In addition, a wide variety of foods are known to be allergenic for certain individuals, but not for the rest of the population, and are, nevertheless, commercialised.

To the best of my knowledge, only Canada requires pre-market risk assessment both for consumers and the environment for all novel plants, regardless of the technology used to improve them (National Academy of Sciences, Engineering and Medicine, 2016). 'Novel foods' refers to food products that have not been eaten in the country and thus do not have a history of safe consumption. Health Canada considers that genetic modifications, in addition to recombinant DNA technologies, are also introduced into the genome of a plant by conventional techniques such as breeding, induced mutagenesis, and the recent gene-editing techniques. Authoritative scientific bodies like the National Academies of the USA state: '*The committee recommends that new varieties – whether genetically engineered or conventionally bred – be subjected to safety testing if they have novel intended or unintended characteristics with potential hazards*' (National Academies of Sciences, Engineering and Medicine, 2016).

Directive 2001/18 establishes that for a GM crop to be authorised in the EU (whether it be for food, feed, as source of ingredients, or for cultivation), it must be subjected to a rigorous pre-market evaluation of potential risks, from two perspectives: consumers' health and the environment. In other words, the agent seeking the commercialisation of that crop must show that it will not cause negative health effects to consumers, nor will it have a negative environmental impact. The current EU

[29] To protect themselves from herbivores, plants synthesize a wide variety of toxic compounds. During plant domestication, through trial and error, humans have (unknowingly) introduced the necessary genetic changes to reduce levels of these compounds so that those plants can now be eaten.

legislation regarding the evaluation of GM crop plants does not address anywhere that these plants could offer benefits nor that these benefits should also be evaluated and taken into consideration when making a decision about their authorisation.

The European Food Safety Authority (EFSA) considers that plants developed through cisgenesis or intragenesis are 'genetically modified' and, therefore, subject to the same authorisation process as other GM plants (EFSA, 2012). In spite of its recognising that hazards associated with cisgenic plants are similar to those of conventionally bred plants, these crops have to be subjected to premarket evaluation, most likely because they fulfil the criteria of the legal definition of 'genetically modified' organisms given in Directive 2001/18. Plants developed by intragenesis are considered to be similar to any GM plant. As of today, only one intragenic crop species has been approved in the USA: the Innateä potato developed by JR Simplot Plant Sciences (JR Simplot Plant Sciences, 2016). It is not authorised in the EU.

Directive 2001/18 does not address 'genome editing techniques' because at that time these techniques had not been developed yet. As mentioned above, in many cases the modification introduced by them cannot be differentiated from modifications introduced by conventional techniques, such as induced mutagenesis, posing, as of yet, unresolved regulatory consequences (Jones, 2015), at least in the EU. In February 2016, the United States Department of Agriculture (USDA) granted non-regulated status (that is, conventional) to a mushroom that was gene-edited to prevent browning, being the first crop developed using CRISPR-Cas9 techniques to be considered not-regulated (USDA-APHIS, 2016) and, therefore, pre-market evaluation was not required for commercialisation, as for any conventionally-bred crop. A herbicide tolerant (HT) canola and other crops developed through gene-editing techniques are expected to be authorised in Canada for the 2017 planting season (CIBUS, 2016).

5.3 Commercialised GM plants in the EU

As of July, 2016, the EU authorises a total of 69 GM varieties belonging to six species (Tables 5.1A and 1B; *EU Register of Genetically Modified Organisms* only for food and feed). Of these, 64 varieties of 5 different crops (cotton, maize, oilseed rape, soybean and sugarbeet; Table 5.1A) can be imported to be used as food and feed, to obtain ingredients to be used in foods and feeds, and other non-food/feed uses. Only one crop plant, a maize variety resistant to the insect European corn borer (MON 810), is authorised for cultivation. Four varieties of carnations (Table 5.1B) can be imported into the EU as cut flowers but are not authorised to be grown in the EU.

A limited variety of traits are currently authorised to be imported as food or feed, or to obtain food ingredients:

- Insect resistance (IR). The plant synthesises a bacterial protein that is toxic for very specific species of insect pests, but does not harm other beneficial insects, other invertebrates, or vertebrates (including humans). There are a total of 8 IR events (an event is defined by the sequence of DNA construct inserted into the plant and the place in the plant genome where it became inserted) in cotton, maize and soybean, of which only one (MON 810) can be grown in the EU.
- Herbicide tolerance (HR). Tolerance to glyphosate is the most frequently used trait. Other herbicides to which tolerance has been engineered into authorised crops are glufosinate

ammonium and dicamba. Some crops are tolerant to two herbicides. In total, there are 19 HR events.
- Drought tolerance is engineered in only one variety of maize.
- Modified oil composition with healthier fatty acids. Stearidonic acid in soybean oil can be converted in the human body to the long chain omega-3 fatty acid eicosapentaenoic acid (EPA). The oil of two varieties of soybean (combined with herbicide tolerance) has a very high concentration of oleic acid (similar to, or somewhat higher than, that in olive oil).
- Male sterility is engineered into a variety of oilseed rape to facilitate the production of hybrids. It cannot be grown in the EU, so these hybrids cannot be produced in the EU.
- Stacked events. Most frequently, an insect resistant plant is crossed with a herbicide tolerant plant to obtain a plant with both traits. These plants are referred to 'stacked events'. Combinations of 3 or 4 different traits are also authorised.
- Modified flower colour is a carnation trait exclusively.

Reflecting on Table 5.1, it is not at all clear why only one GM crop can be cultivated in the EU. Although I am not aware of any reason (scientific or otherwise) that may have been given to account for this fact, three possible explanations can be put forward:
- The interested agent(s) did not apply for an authorisation to grow other crops considering the very slow and irregular pace of approvals of genetically engineered (GE) crops in the EU (Aldemita *et al.*, 2015; EuropaBio, 2014).
- The interested agent(s) did not apply for cultivation to save the time and cost of gathering the data for the required environmental impact statement related to the crop of interest.
- Authorisation may not be granted to avoid gene transfer from the GM crop to related wild species in the cases for which those wild species exist in Europe.

As mentioned above, the authorised maize variety carries only the trait conferring resistance to the insect European corn borer. No other insect resistant crops can be grown in the EU, not even those which do not carry any other HT stacked traits. Only one maize variety out of five insect resistant varieties can be grown. All insect resistant genes inserted in the crop varieties authorised for import come from the bacterium *Bacillus thuringiensis*, being the Cry genes inserted in 7 cases and the vip3A gene inserted in only one case. Insecticidal Bt genes (as they are collectively referred to) have been very well studied and characterised (Niederhuber, 2015), including the most recently used one, vip3A (Estruch *et al.,* 1996). In fact, formulations of the whole bacterium *B. thuringiensis* (carrying all its genes!) have been widely used since the middle of the 20th century and are authorised for insect control in organic agriculture (European Council, 2007). Insect resistant cotton could be very useful in Southern Spain where the 2015 harvest decreased, on average, between 15 and 20% with respect to the previous year, down to between 43 and 48% in the most heavily infested areas (ASAJA, 2015; USDA, 2015). Subsidies did play an important role in planting decisions.

No GE herbicide tolerant plant is authorised for cultivation in the EU for no apparent reason. As mentioned above, we could speculate that perhaps the reason is to avoid accidental hybridisation with some of the wild species of oilseed rape and sugarbeet present in Europe which could convert them in difficult to manage weeds, albeit only under certain circumstances. However, in Europe there are no wild relatives of maize, cotton or soybeans because the centres of origin for these crops are in Mexico, several subtropical regions, and China, respectively (Smartt and Simmonds, 1995).

Surprisingly, Romania grew herbicide tolerant soybeans between 2001 and 2006 (James, 2006) with no apparent problems either in Romania or in other EU countries, but had to stop growing this crop in 2007 when it became part of the EU.

It should be noted that all currently grown conventional crops (improved by non-GM techniques) tolerate at least one herbicide, if not more than one. See, for example, the chart of herbicides that can be used with corn and soybean (Pioneer, 2016). A recent example of a conventional technique to introduce the herbicide tolerance characteristic is the ClearfieldÒ technology developed by BASF (BASF, 2016). Using mutation breeding, BASF developed the first non-GM wheat which could grow well in the presence of imidazolinone herbicide (Health Canada, 2007). This technology has been used to develop non-GM canola, lentils, sunflower and rice. In addition to the seeds, BASF also commercialises the herbicide tolerated by its seeds, as can be seen in these web pages. Actually, the practice of patenting and commercialising a given product and all the other products the former one needs to function is very frequent in all industrial sectors of today's global world, with Apple being but one of the widest known examples.

Knowing that there are no wild relatives of soybean in Europe, it is difficult to understand why soybeans with healthier fatty acids cannot be cultivated. Similarly, drought tolerant maize, which could be very appropriate for Southern regions of Europe where the rise in temperatures is creating variable weather patterns with more frequent drought periods. This crop can be imported but not cultivated on EU soil.

For unknown reasons, the EU does not appear to consider appropriate the cultivation of any other GM crops which are currently authorised for import, *de facto* blocking cultivation of GM crops. Several studies indicate that in countries where GM crops can be grown, farmers have substantial economic benefits from increased yields and reduced expenses due to inputs and easier crop management (Brookes and Barfoot, 2016; Taheripour *et al.,* 2015). If the limited experience of Spain with Bt maize is an indication (Riesgo, 2014), farmers in other EU countries would certainly benefit from growing GM crops.

5.4 Authorisation pending GM crops and trade constraints

Another 22 GM events of a total of 5 crops have received positive reports from the EFSA but are still not authorised to be imported into the EU (Table 5.2): 8 maize varieties, 8 soybean varieties, 3 rapeseed varieties, 2 cotton varieties and 1 rice variety. All of them have been authorised in other countries. This difference in the timing of authorisation of GM crops in the EU and in other producing countries from which the EU may import GM crops is known as 'asynchronous approval' (Henseler *et al.,* 2013).

Asynchronous approval has important implications for international trade. In the EU, the livestock industry is the one primarily affected, among various economic sectors. For example, during 2013/14 the EU imported approximately 34 million ton of soybeans (includes soymeal; representing 96% of the soymeal consumed in the EU (Futtermittel, 2016) from the United States, Brazil and Argentina, the leading countries for soybean export for livestock production. These countries in addition to growing the events currently authorised in the EU, they also grow others for which authorisations are pending in the EU and, consequently, cannot be imported. Therefore, for producers in non-EU countries to be able to export to the EU, it is mandatory to ensure costly segregation procedures, from the field throughout all steps of the supply chains and all the way to the end customer. In addition, expensive testing methodologies need to be set up at critical points of the supply chain to ascertain that no unauthorised events are present in the export shipments, thus avoiding the high cost of stopping shipments at EU ports if traces of unauthorised events are detected. All these measures inevitably result in price increases (Kalaitzandonakes *et al.*, 2014). Alternatively, if the segregated supply chain and its corresponding testing methodologies are not economically feasible, trade of the product between the two countries may cease (National Academies of Sciences, Engineering and Medicine, 2016). Considering that GM soybeans in 2015 were grown in 11 non-EU countries and that about 83% of the soybeans globally produced in 2014 are genetically modified (James, 2016), if imports of GM soybeans were restricted, sourcing non-GM soybeans would not be easy and would certainly lead to substantial price increases for this very important commodity.

On January 15, 2016, the European Ombudsman, in response to the complaints presented by industry associations Association of European Biotechnology Industries (EuropaBio), European Confederation of National Associations trading in cereals and feedstuffs (COCERAL) and European Feed Manufacturers' Federation (FEFAC), *'found that the delays affecting the twenty applications were not justified. In the course of the inquiry, the Commission dealt with all the pending applications. The Ombudsman concluded, however, that the delays reflected a systemic problem rather than being the result of matters specific to the particular authorisation applications. In closing the inquiry, the Ombudsman found that the delays constituted maladministration on the part of the Commission' (European Ombudsman, 2016). On* June 9, 2016, considering that the situation had not changed, trade Associations COCERAL, FEDIOL (Federation of European Vegetable Oil and Proteinmeal Industries) and FEFAC addressed European Commission President Jean-Claude Juncker a letter (COCERAL *et al.*, 2016) requesting the import authorisation of GM soybeans MON87705 x MON89788, MON87708 x MON89788 and FG72 to ensure the viability of feed supplies into the EU.

Soybeans are the main protein source for feed materials, representing 70% of the protein required (Baulcombe *et al.*, 2014). As stated above, the EU currently imports 95-96% of the total amount of soybeans it needs and it would require an acreage of 13.5 million hectares to be self-sufficient instead of the 0.6 million hectares it currently cultivates. In light of all the above, it is difficult to understand why GM soybeans cannot be cultivated in the EU.

A study commissioned by the Directorate-General for Agriculture and Rural Development of the European Commission concluded that if the EU would not be able to import soybeans from those countries due to the 'asynchronous approval' problem, the price of soybean products for feed could increase by over 200% in the next couple of years. The strong livestock industry which represents 40% of the agricultural production in the EU would be in serious danger of decreasing substantially because it would be difficult for farmers to produce the necessary raw materials for feeds in a short time. 'Profits could decrease by € 1.2 billion for the dairy sector, by € 3 billion for the beef sector, by € 1 billion for the pork meat sector, and by € 380 million for the poultry meat sector' (Agricultural Economics Research Institute, 2010). Thus, to avoid damaging its own livestock and related industries, the European Commission approved Regulation 619/2011 (EC, 2011) establishing the definition (and methods of detection) of 'technical zero' level for detection of asynchronous and obsolete events in feed only. A 'technical zero' is set at the level of 0.1%, the lowest level which is satisfactorily reproducible among laboratories (EC, 2016c). Yet, for food applications, the 'zero-tolerance' policy still applies. Segregating the supply chain to this level of as-yet non-authorised events is truly unrealistic. What is the practical use of having a 'zero-tolerance' level for as-of-yet unapproved events for food applications in the EU if that level cannot be satisfactorily reproduced among laboratories, as agreed to by the European Commission? It certainly does not make sense.

The cost of asynchronous approvals and the subsequent problems for the livestock industry would certainly affect European consumers through price increases for foods of animal origin (meat, eggs and dairy products). Livestock farmers would likely try to pass on to consumers most of the added feed costs which are expected to be as high as € 9.6 billion (Agricultural Economics Research Institute, 2010). If this is not possible, farmers will have to assume these costs with the concomitant result of a severe loss of competitiveness for the sector. If finally costs are passed on to consumers, an increasing number of the European population will not be able to afford a meal with meat, or chicken, every second day. Currently, in the EU (28 countries) this number is, on the average, 9.0% of the population (Eurostat, 2016). As a result, an increase in malnutrition and its health and societal associated costs could be expected to increase in the near future.

As a comparison, and to underline the absolute lack of scientific logic behind the (technical) 'zero tolerance' level, two outstanding examples are worth mentioning. Council Regulation 834/2007 on organic production accepts an unspecified level of 'adventitious presence' of unauthorised synthetic compounds in organic crops and up to 5% non-organic ingredients in processed organic foods (European Council, 2007). In 2007, the EU increased the maximum acceptable levels of mycotoxins (some of which are known carcinogens) in maize and maize products to 'avoid a disruption in the market' (see 'Whereas Clause 6 of Commission Regulation 1126/2007) (EC, 2007) because in those years conventional maize and derived processed products had excessively large levels of these compounds. In 2013, because the level of mycotoxins in its maize and maize products were much higher than the newly-raised levels, France requested that the levels of mycotoxins in these products be temporarily suspended to avoid a disruption of the market. EFSA concluded that, if temporarily suspended, these new levels of these toxic compounds presented a health concern particularly for children (EFSA, 2014).

The result of the administrative malpractice regarding GM crops is an important reduction in research and development of new products for the EU market, and the relocation of a substantial number of highly qualified jobs to other non-EU countries. In addition, the research actually conducted in the EU is, most often, is focused on crops for other non-EU countries (Späth, 2016)

5.5 Factors affecting low level presence

The term 'low level presence' refers to the low concentrations (variable, although often trace amounts) of a given GM crop caused by the asynchronous approval of these crops in different countries (FAO, 2014) which cause regulatory and economic problems. In addition to the time difference in approval of GM crops among countries, a combination of several factors of various types, some of which may be unavoidable, is responsible for this 'low level presence':

- ease of detection of transgenic elements;
- biology of the various plants involved;
- types of approved GM crops.

It is important to note that the genetic modifications introduced by genetic engineering techniques can be easily detected by the very sensitive polymerase chain reaction (PCR) techniques (Ahmed, 2004). Thus, very small amounts of certain DNA sequences which are part of the inserted genes can be detected – and quantified – in the GM crop of interest. In contrast, genetic modifications made with any of the conventional techniques, particularly by the recent genome editing techniques, are much more difficult to detect, if not practically impossible in the large majority of cases, unless the modification introduces an easily observed phenotypic characteristic. For example, the plant now tolerates an herbicide which previously killed it, as it occurs with ClearfieldÒ wheat as mentioned above. Thus, all GM plants, but not conventional ones, carry 'a red flag' waiting to be detected.

The ease of detection of GM elements is at the heart of the issues of the social debate and the regulatory procedures surrounding the GM technologies because the presence of GM elements from a GM plant can be detected in a conventional plant.

A common characteristic of all the grains, fleshy fruits and vegetables that we eat and the cotton and flax that give us fibre for our clothes is that they all have flowers. Throughout evolution, flowering plants have developed many mechanisms to ensure that the pollen is delivered to the stigma of a flower, whether it is a different flower in a different plant (cross-pollination) or the same flower (self-pollination). For example, maize and other grasses are wind pollinated, whereas soybeans are self-pollinated to a large extent, and rapeseed may be pollinated either by wind or bees (Cubero, 2013). Gene transfer among closely related plants is of critical importance in evolution and has taken place in nature since the time plants with flowers (angiosperms) evolved (Raven *et al.,* 1999). It is certainly not 'genetic contamination', unless genetic contamination is considered as a major evolutionary mechanism. For wind pollinated maize, detailed studies have been conducted to determine the practical distance between GM and conventional maize to maintain cross-pollination below certain threshold levels (Riesgo *et al.,* 2010).

GM crops can be approved for different specific purposes. Thus, an experimental variety (event) can be approved only for field trials, whether confined or on open fields, and, if the proper care is not exercised, it may end up, partially or totally, mixed with the conventional crop (EFSA, 2005; Pollack, 2002). A second situation is that of a GM variety approved only for a given application (for example, feed) but not for others (i.e. food, and food ingredients) and, if segregation along the production chain is not efficient, traces can be detected in the non-approved application. This is the case of the Starlink maize which was approved only for feed applications but not for food (Bucchini *et al.,* 2002). As a result, no other 'split' applications for commercialisation (that is, only for feed, and not for feed and food) have been presented. A third case is that of a GM variety approved only for a given period of time which may still appear at very low concentrations for a variety of reasons (e.g. presence of volunteers in the field in following years). Finally, as described under 'asynchronous approval', a GM variety is approved for commercial uses in some countries but not in others.

In all these cases, consumers' health is most likely not compromised, nor are the unapproved events likely to pose environmental problems different from those posed by plants improved through non-gm techniques. The agent seeking authorisation applies for a specific application (field trials in open fields, to commercialise the given event only for feed, etc.) submitting the required data.

5.6 GM crops, human and animal health, the environment and the EU consumer

Because we are dealing with crops destined to be eaten, both by people and by animals which provide us with food, and must be grown in open fields, legislation of most countries where these crops are commercialised requires that they do not cause harm either to people and animals or to the environment.

In the Executive Summary of its recent report, the Committee on Genetically Engineered Crops of the National Academies of Sciences, Engineering and Medicine of the United States clearly affirms that '*long-term data on livestock health before and after the introduction of genetically engineered crops showed no adverse effects associated with Genetically Engineered crops. The committee also examined epidemiological data on incidence of cancers and other human-health problems over time and found no substantiated evidence that foods from GE crops were less safe than foods from non-GE crops.*' (National Academies of Sciences, Engineering and Medicine, 2016). The European Commission itself accepts that 'the main conclusion that can be drawn from over 130 research projects carried out during the last 25 years by over 500 independent research groups is that biotechnology and particularly GM organisms are not *per se* more dangerous than conventional techniques for plant genetic improvement' (EC, 2014). Thus, from the point of view of animals and consumers' health, GM crops and the food and feed derived from them are, at least, as safe – or unsafe – as their non-GM counterparts. Yet, in some cases, GM crops can be safer than their non-GM counterparts.

Some GM crops with reduced toxin levels are safer for consumers than their non-GM counterparts. The only GM crop authorised for cultivation in the EU, maize resistant to the European corn borer, has considerably lower levels of fumonisins, toxic compounds produced by the fungus *Fusarium* when it infects maize plants through the 'wounds' produced by insects feeding on the plant (Bakan *et al.,* 2002; Wu, 2006). Fumonisins interfere with the cellular absorption of folic acid, an essential vitamin, particularly during foetal development to prevent serious congenital malformations of the nervous system such as spina bifida. The presence of mycotoxins in maize and derived products has been linked to these congenital malformations (Etzel, 2002). Fumonisins and other mycotoxins are also present in organically grown maize (Ariño *et al.,* 2007; Rommens *et al.,* 2008). The second example, although to date only authorised in the US, is a potato which produces very low levels of acrylamide upon frying (APHIS, 2014). Acrylamide, a neurotoxin and cancer causing chemical, is spontaneously produced when frying potatoes (US Food and Drug Administration, 2016). In addition, this potato does not develop brown spots, thus substantially cutting down waste. Although an enzymatic solution for decreasing the level of acrylamide upon frying is available (Novozymes, 2012), the enzyme is not available at the retail level, whereas fresh potatoes, if approved in the EU, would be accessible to consumers. EU consumers, as well as potato growing farmers, would greatly benefit from it, if it were authorised in the EU. A Spanish research group has developed a GM wheat with 97% less gliadin than normal wheat, the flour of which can be used to make bread acceptable for celiac sufferers (Gil-Humanes *et al.*, 2014). Although still at the testing stage, it is hoped that this GM crop, developed with EU public money in a public research institute, will benefit European celiacs in the short-medium term (Laursen, 2016).

A currently authorised soybean (event MON 87769-7) accumulates stearidonic acid in its seeds, yielding an oil rich in this fatty acid which is converted in the body to the 'fish' omega-3 fatty acid, EPA (Ursin, 2003). A crop capable of yielding oil enriched in the EPA fatty acid is the oil plant *Camelina sativa,* related to the rapeseed, still at the field trials stage in the UK (Usher *et al.,* 2015). These two oils will provide consumers with a sustainable plant source of these very necessary fatty acids without the need to eat fish.

Various staple GM crops (rice, 'Golden Rice'[30]; corn (Naqvi *et al.,* 2009); bananas (Queensland University of Technology, 2014) with enriched levels of vitamins (mostly vitamin A) have been developed to help reduce the significant deficit of this nutrient, which causes severe health problems in developing countries among the population who does not have access to a balanced diet including fruits and vegetables. Golden Rice and the corn have been developed in Europe with EU public funds whereas the banana was developed in Australia. These biofortified crops may be of little use to European consumers who can afford a variety of nutritious foods in the desired amounts. The opposition of certain groups, as well as the negative European attitude towards GM crops, is causing that the rice, free from patent royalties and available for planting for the last few years, has not yet been distributed among subsistence farmers for planting[31]. On June 30, 2016, 109 Nobel Prize winners signed a letter addressed to Greenpeace, the United Nations and Governments around the world, urging the first one to stop its opposition to Golden Rice (Laureates, 2016). This letter is very significant because both the vitamin A-rich banana and the hypervitaminic corn, as well as other biofortified crops, may end up suffering the same fate.

The greatest majority of GM crops in the European Union is dedicated to animal feeds because of the public's rejection for food. The food industry is reluctant to use ingredients derived from GM crops into processed foods because the EU labelling regulations (European Council, 2003) require labelling if the amount of a given ingredient is equal to, or higher than, 0.9% of the total amount of that ingredient present in the food, even if detection in the food is practically impossible (i.e. refined vegetable oils, sugar or high fructose corn syrup). For the European consumer the main benefit of the current situation is that, thanks to the imports of GM soy products for feed, prices of animal-derived foods will remain affordable if the EU internal market is not disrupted, as described above.

GM crops do offer a variety of environmental advantages, the extent of which will depend on the particular circumstances and characteristics of each farm. The report of the National Academies of Sciences, Engineering and Medicine already mentioned, dedicates a long chapter (Chapter 4) to examine the available evidence related to agronomic and environmental effects of GE crops (National Academies of Sciences, Engineering and Medicine, 2016). Insect resistant crops '*generally decreased yield losses and the use of insecticides on small and large farms in comparison with non-Bt varieties*'.

[30] Golden Rice: http://www.goldenrice.org.

[31] Allow Golden Rice now: http://www.allowgoldenricenow.org.

Both in the US and in China, statistical evidence over several years indicates that the populations of insect pests in maize and cotton fields has been substantially reduced, for the benefit of all farmers, adopters and non-adopters of Bt crops. Experience with Bt maize in Spain indicated that average yield enhancement, although varied depending on local circumstances, could get as high as 12% without added inputs (Park *et al.,* 2011). In studies carried out in Spain between 1998 and 2015 comparing Bt maize varieties with their conventional isogenic counterparts (grown in the same field and under identical conditions) yield losses of conventional varieties can get as high as 30 or 40% under severe corn borer infestations which may represent economic losses higher than 250 €/ha, considering an average yield of 14 Tm/ha (Tribó *et al.,* 2016).

An average reduction in pesticide use of 36.9% was reported in a meta-analysis of 147 recent publications regarding soya, maize and cotton (Klümper *et al.,* 2014). Higher insect biodiversity has been reported in fields with Bt crops than in those with conventional crops treated with synthetic insecticides which are not nearly as species-specific as the Bt toxins are. The increase in yield for HT crops occurs mainly where weed control is improved.

HT crops use primarily glyphosate and glufosinate ammonium as herbicides, both of which are low-environmental impact herbicides. The use of HT crops decreased the use of more aggressive herbicides (National Academy of Sciences, 2010), resulting in lower environmental impact quotients for HT soya than those for conventional soya (Brookes, 2014; Fernández-Cornejo *et al.,* 2014). HT crops greatly facilitate crop management and the implementation of reduced tillage or no-till (Fernández-Cornejo *et al.,* 2012) conservation agriculture. The not-so-positive aspect of glyphosate-tolerant crops is that improper and excessive use of glyphosate, without adequate stewardship, is causing the appearance of glyphosate resistant weeds (Duke and Powles, 2009) which may jeopardise the use of this technology. Control of glyphosate resistant weeds is achieved with crop rotations and the use of other herbicides (University of Wisconsin Extension Services, 2004), usually of greater environmental impact than that of glyphosate, which in turn is causing the environmental impact quotient of certain HT crops to increase in the last few years (Fernández-Cornejo *et al.,* 2014).

In March 2015 the International Agency for Research on Cancer (IARC) of the World Health Organisation reclassified glyphosate from 'possibly carcinogenic to humans' to 'probably carcinogenic to humans' (Guyton *et al.,* 2015). This declaration sparked a controversy, particularly due to neglecting the fact that the IARC does not perform risk assessment but only defines hazards. It should be pointed out that a hazard is '*any source of potential damage, harm or adverse health effects on someone*', whereas risk is '*the chance or probability that a person will be harmed or experience an adverse health effect if exposed to the hazard*' (Canadian Center for Occupational Health and Safety, 2016). The current list of agents classified by the IARC Monographs includes in the same group as glyphosate emissions from high-temperature frying, occupational exposure as hairdresser or barber, drinking very hot beverages above 65 °C, red and processed meat, or shiftwork that involves circadian disruption, among a number of specific chemical products (IARC, 2016). In November 2015, EFSA concluded that 'glyphosate is unlikely to pose a carcinogenic hazard to humans and the evidence does not support classification with regard to its carcinogenic potential according to Regulation (EC) No 1272/2008' (EFSA, 2015). In May 2016, the Joint FAO/WHO Meeting on Pesticide Residues says that '*it is notable that the only cohort study of high quality found no evidence of an association at any exposure level*', and it concludes that '*glyphosate is unlike to pose a carcinogenic risk to humans from exposure through the diet*' (FAO/WHO, 2016).

A recent review of articles published between 2002 and 2012, with 847 articles dealing with environmental issues, found no significant risks related to planting and growing GM crops as compared to those posed by conventionally-bred crops (Nicolia *et al.,* 2013).

A concomitant benefit of the increase in conservation agriculture is an important reduction in CO_2 emissions and fuel consumption through fewer applications of phytochemicals. The increase in productivity in the 4 GM major crops (maize, soybeans, cotton and canola) during 2015 'saved' approximately 20 million ha which would have been needed if these crops had not been GM (Brookes and Barfoot, 2016; James, 2015). Conservation agriculture also reduces greenhouse gas emissions from the soil in significant amounts (Mangalassery *et al.,* 2014).

Drought tolerant maize, currently allowed to be imported into the EU but not cultivated, can maintain its usual yield under seasonal drought conditions. It is grown only in the US and during 2015 it was planted on 810,000 ha, a 3-fold increase from 2014, implying good farmers' acceptance (James, 2015). Southern Europe would certainly benefit from it.

To capitalise on the many positive environmental features of HT crops, and to achieve a sustainable use of both Bt and HT crops, integrated pest-management strategies need to be implemented (National Academies of Sciences, Engineering and Medicine, 2016).

In a nutshell, GM crops can offer many benefits to European consumers and farmers, to Europe's environment and economics. GM technology and the recent gene-editing technology are the latest additions to the varied palette of techniques that we currently can use to improve the crops we eat and to lessen the environmental impact of agriculture. They should be used when and where appropriate to help farmers and consumers to produce sufficient nutritious food and the lower the burden of agriculture on the environment. FAO Director General, José Graziano da Silva, called for a 'paradigm shift'

> to lower the use of agricultural inputs, specially water and chemicals to put agriculture... on a more sustainable and productive long-term path. Options such as agro-ecology and climate-smart agriculture should be explored, and so should biotechnology and the use of genetically modified organisms... We need to explore these alternatives using an inclusive approach based on science and evidences, not on ideologies (Da Silva, 2014).

Table 1A. Genetically modified food and feed in the EU.[1]

Crop	Company	Unique identifier and trait(s)	Authorisation exp. date
Cotton	Monsanto	MON 01445–2 (HT)[2]	26/4/2025
	Monsanto	MON 15985–7 (IR)[3]	26/4/2025
	Monsanto	MON 00531–6 (IR)	26/4/2025
	Monsanto	MON 00531–6 × MON 01445–2 (IR+HT)	26/4/2025
	Monsanto	MON–88913–8 (HT)	26/4/2025
	Bayer	ACS–GHØØ1–3 (HT)	28/10/2018
	Bayer	BCS–GH 002–5 (HT)[a]	16/6/2021
	Bayer	BCS–GHØØ4–7 (IR+HT)	26/4/2025
	Bayer	BCS–GHØØ2–5×ACS–GHØØ1–3 (HT)	26/4/2025
	Dow Agro-Sciences	DAS–24236–5 (×DAS–21Ø23–5 (HT+IR)	21/12/2021
Maize	Syngenta	SYN–BT Ø11–1 (IR)	27/7/2020
	Syngenta	MON–ØØØ21–9 (HT)	27/3/2018
	Syngenta	MON–ØØØ21–9 (HT)	27/3/2018
	Syngenta	SYN–IR6Ø4–5 (IR)	29/11/2019
	Syngenta	SYN–BTØ11–1×MON–ØØØ21–9 (IR+HT)	27/7/2020
	Syngenta	SYN–IR6Ø4–5×MON–ØØØ21–9 (IR+HT)	21/12/2021
	Syngenta	SYN–BTØ11–1×SYN–IR6Ø4–5 (IR+HT)	21/12/2021
	Syngenta	SYN–BTØ11–1×SYN–IR6Ø4–5×MON–ØØØ21–9 (IR+HT)	21/12/2021
	Syngenta	SYN–IR162–4 (IR)	18/10/2022
	Pioneer & Dow AgroSciences	DAS–Ø15Ø7–1×MON–ØØ6Ø3–6 (IR+HT)	23/10/2017
	Pioneer & Dow AgroSciences	DAS–59122–7 (IR+HT)	23/10/2017
	Pioneer & Dow AgroSciences	DAS–Ø15Ø7–1 (IR+HT)	Renewal ongoing
	Monsanto	MON–ØØ81Ø–6 (IR)[4]	Pollen: 5/11/2023 Others: renewal ongoing
	Monsanto	MON–ØØ6Ø3–6 (HT)	26/4/2025
	Monsanto	MON–ØØ6Ø3–6×MON–ØØ81Ø–6 (HT+IR)	23/10/2017
	Monsanto	MON–88Ø17–3 (IR+HT)	29/10/2019
	Monsanto	MON–89Ø34–3 (IR)	29/10/2019
	Monsanto	MON–88Ø17–3×MON–ØØ81Ø–6 (IR+HT)	27/7/2020
	Monsanto	MON–89Ø34–3×MON–ØØ6Ø3–6 (IR+HT)	27/7/2020
	Monsanto	MON–89Ø34–3×MON–88Ø17–3 (IR+HT)	16/6/2021
	Monsanto	MON–89Ø34–3×DAS–Ø15Ø7–1×MON–88Ø17–3×DAS–59122–7 (IR+HT)	5/11/2023

Table 1A. Continued.

Crop	Company	Unique identifier and trait(s)	Authorisation exp. date
	Monsanto	MON–89Ø34–3×DAS–Ø15Ø7–1×MON–88Ø17–3 (IR+HT)	5/11/2023
Maize	Monsanto	MON–89Ø34–3×DAS–Ø15Ø7–1×DAS–59122–7 (IR+HT)	5/11/2023
	Monsanto	MON–89Ø34–3×MON–88Ø17–3×DAS–59122–7 (IR+HT)	5/11/2023
	Monsanto	DAS–Ø15Ø7–1×MON–88Ø17–3×DAS–59122–7 (IR+HT)	5/11/2023
	Monsanto	MON 8746Ø–4 (drought tolerant)	26/4/2025
	Monsanto	MON–ØØ6Ø3–6×ACS–ZMØØ3–2 (HT)	3/12/2025
	Monsanto	MON–87427–7 (HT)	3/12/2025
	Monsanto & Dow AgroSciences	MON–89Ø34–3×DAS–Ø15Ø7–1 (IR+HT)	5/11/2023
	Monsanto & Dow AgroSciences	MON–89Ø34–3×DAS–59122–7 (IR+HT)	5/11/2023
	Monsanto & Dow AgroSciences	DAS–Ø15Ø7–1×MON–88Ø17–3 (IR+HT)	5/11/2023
	Monsanto & Dow AgroSciences	MON–88Ø17–3×DAS–59122–7 (IR+HT)	5/11/2023
	Monsanto & Dow AgroSciences	MON–89Ø34–3×DAS–Ø15Ø7–1×MON–ØØ6Ø3–6 (IR+HT)	5/11/2023
	Bayer	ACS–ZMØØ3–2 (HT)	26/4/2025
	Pioneer	DAS–59122–7×MON–ØØ6Ø3–6 (IR+HT)	29/10/2019
	Pioneer	DAS–59122–7×DAS–Ø15Ø7×MON–ØØ6Ø3–6 (IR+HT)	27/7/2020
	Pioneer	DAS–Ø15Ø7×DAS–59122–7 (IR+HT)	27/7/2020
Oilseed Rape	Monsanto	MON–ØØØ73–7 (HT)	26/4/2025 Renewal ongoing
	Monsanto	MON–883Ø2–9 (HT)	26/4/2025
	Bayer	ACS–BNØØ5–8ACS–BNØØ3–6ACS–BNØØ5–8×ACS–BN003–6 (HT+male sterility)	26/4/2023
	Bayer	ACS–BNØØ8–2 (HT)	9/3/2019
Soybean	Bayer	ACS–GMØØ5–3 (HT)	7/9/2018
	Bayer	ACS–GMØØ6–4 (HT)	9/2/2022
	Monsanto	MON–89788–1 (HT)	3/12/2018
	Monsanto	MON–Ø4Ø32–6 (HT)	9/2/2022
	Monsanto	MON–877Ø1–2 (IR)	9/2/2022
	Monsanto	MON–877Ø1–2×MON–89788–1 (IR+HT)	27/6/2022
	Monsanto	MON–877Ø5–6 (HT+high oleic acid)	26/4/2025
	Monsanto	MON–877Ø8–9 (HT)	26/4/2025

Table 1A. Continued.

Crop	Company	Unique identifier and trait(s)	Authorisation exp. date
	Monsanto	MON–87769–7 (high stearidonic acid)	26/4/2025
	Pioneer	DP–356Ø43–5 (HT)	9/2/2022
	Pioneer	DP–3Ø5423–1 (HT+high oleic acid)	26/4/2025
	BASF	BPS–CV127–9 (HT)	26/4/2025
Sugarbeet	KWS SAAT and Monsanto	KM–ØØØ71–4 (HT)	23/10/2017

[1] Adapted from the EU Register of authorised GMOs (European Commission, 2016a).
[2] HT = herbicide tolerant (mostly glyphosate; other herbicides: glufosinate ammonium and dicamba; some crops can be tolerant to more than 1 herbicide).
[3] IR = insect resistant (different species of Lepidoptera (mostly) or other orders; some crops can be tolerant to more than 1 insect species).
[4] This is the only crop authorised for cultivation in the EU.

Table 1B. Non-food or feed authorised GM plants in the EU.[1]

Plant	Company	Unique identifier, trait(s) and scope of authorisation	Authorisation exp. date
Carnation	Florigene	IFD–5958–3 HT – import and marketing	26/4/2025
	Florigene	IFD–264Ø7–2 HT – import and marketing	26/4/2025
	Florigene	FLO–4Ø689–6 – modified flower colour – import and marketing	15/3/2019
	Florigene	FLO–4Ø644–4 – modified flower colour – import and marketing	30/5/2017

[1] Adapted from The GMO Compass Database (GMO Compass, 2016).

Table 2. GM crops for which an authorisation is pending.[1]

Name of the product Company	Unique identifier and trait	Date of valid EFSA application or opinion published	Countries in which crop cultivation is authorised[2]
Rice – LLRICE62 Bayer CropScience	ACS–OSØØ2–5 (HT[3])	EFSA Opinion published 30/11/2007	USA
Maize – 3272 Syngenta	SYN–E3272–5 (thermotolerant alpha amylase)	EFSA opinion published 20/06/2013	Canada, Japan, USA
Maize – Bt11×MIR162×MIR604×GA21 Syngenta	SYN–BTØ11–1×SYN–IR162–4×SYN–IR6Ø4–5×MON–ØØØ21–9 (HT+IR[4])	EFSA opinion published 7/12/2015	Argentina, Brazil, Canada
Maize – Bt11×MIR162×1507×GA21 Syngenta	SYN–BTØ11–1×SYN–IR162–4×DAS–Ø15Ø7–1×MON–ØØØ21–9(HT+IR)	Valid EFSA application 14/06/2012	Argentina, Canada, Japan
Maize – DAS–40278–9 Dow AgroSciences	DAS–4Ø278–9 (HT)	Valid EFSA application 11/03/2011	Brazil, Canada, USA
Maize – 1507×59122×MON 810×NK603 Pioneer	DAS–Ø15Ø7–1×DAS–59122–7×MON–ØØ81Ø–6×MON–ØØ603–6 (HT+IR)	Valid EFSA application 30/01/2012	Canada, Japan
Maize – 5307 Syngenta	SYN–Ø53Ø7–1 (IR)	EFSA opinion published 05/05/2015	Canada, Japan, USA
Maize – Bt11×MIR162×MIR604×1507×5307×GA21 Syngenta	SYN–BTØ11–1×SYN–IR162–4×SYN–IR6Ø4–5×DAS–Ø15Ø7–1×SYN–Ø53Ø7–1×MON–ØØØ21–9 (HT+IR)	Valid EFSA application 18/08/2014	Canada
Maize – Bt11×59122×MIR604×1507×GA21 Syngenta	SYN–BTØ11–1×DAS–59122–7×SYN–IR6Ø4–5×DAS–Ø15Ø7–1×MON–ØØØ21–9 (HT+IR)	Valid EFSA application 14/06/2012	Canada, Japan
Soybean – 305423×40–3–2 Pioneer	DP–305423–1×MON–04032–6	Valid EFSA application 19/02/2008	Argentina, Canada, Japan
Soybean – MON 87769×MON 89788 Monsanto	MON–87769–7×MON–89788–1 (HT+Modified oil fatty acids)	EFSA Opinion published 8/10/2015	USA, Mexico, South Korea, Taiwan
Soybean – DAS–68416–4 Dow AgroSciences	DAS–68416–4 (HT)	Valid EFSA application 08/09/2011	Brazil, Canada, Japan, USA

Table 2. Continued.

Name of the product Company	Unique identifier and trait	Date of valid EFSA application or opinion published	Countries in which crop cultivation is authorised[2]
Soybean – FG72 – Bayer	MST-FGØ72-2 (HT)	EFSA opinion published 16/07/2015	Canada, Japan, USA
Soybean – MON 87705×MON 89788 Monsanto	MON-877Ø5-6×MON-89788-1 (HT+Modified oil fatty acids)	EFSA Opinion published 16/07/2015	Mexico, South Korea, Taiwan
Soybean – DAS-44406-6 Dow AgroSciences	DAS-444Ø6-6 (HT)	Valid EFSA application 15/04/2013	Argentina, Canada, Japan, USA
Soybean – MON 87708×MON 89788 Monsanto	MON-877Ø8-9×MON-89788-1 (HT)	EFSA opinion published 18/06/2015	Colombia, Japan, Uruguay
Soybean – DAS-81419-2 Dow AgroSciences	DAS-81419-2 (HT+IR)	Valid EFSA application 7/02/2014	Canada, Japan, USA
Rapeseed – MS8×RF3×GT73 Bayer	ACS-BNØØ5-8×ACSBNØØ3-6×MON-ØØØ73-7 (HT+Male sterility+Fertility restoration)	Valid EFSA application 11/05/2012	Japan
Oilseed rape – 73496 Pioneer	DP-073496-4 (HT)	Valid EFSA application 4/12/2012	Australia, Canada, Japan, USA
Oilseed rape – MON 88302×Ms8×Rf3-Monsanto	MON-883Ø2-9×ACSBNØØ5-8×ACS-BNØØ3-6 (HT+Male sterility+Fertility restoration)	Valid EFSA application 24/04/2014	Japan
Cotton – GHB119-Bayer	BCS-GHØØ5-8 (HT+IR)	Valid EFSA application 21/11/2011	Japan, USA
Cotton – 281-24-236×3006-210-23×MON 88913 Monsanto & Dow AgroScience	DAS-21023-5×DAS-24236-5×MON-88913-8 (HT+IR)	EFSA Opinion Published 8/04/2016	Costa Rica, Japan

[1] Adapted from the EU Register of authorised GMOs (European Commission, 2016a).
[2] Data in this column is taken from the International Service for the Acquisition of Agri-Biotech Applications GM Approval Database (ISAAA GM Approval Database, 2016).
[3] HT = herbicide tolerance. The crop may be tolerant to one or more herbicides.
[4] IR = insect resistance. The crop may be resistant to one or more insect pests.

References

Agricultural Economics Research Institute (2010). Study on the implications of asynchronous GMO approvals for EU imports for animal feed products: executive summary. Contract No. 30-CE-0317175/00-74. Available at: http://tinyurl.com/znvp6sk.

Ahmed, F.E. (2004). Testing of Genetically Modified Organisms in food. Haworth Press, Philadelphia, PA, USA.

Akeley, R.V., Mills, W.R., Cunningham, C.E. and Watts, J. (1968). Lenape: a new potato variety high in solids and chipping quality. American Potato Journal 45: 142-145.

Aldemita, R.R., Reaño, I.M., Solís, R.O. and Hautea, R.A. (2015). Trends in global approvals of biotech crops (1992-2014). GM Crops and Foods 6: 150-166.

Álvares, R.L. (2014). 15 Años de cultivo de maíz Bt en España: beneficios económicos, sociales y ambientales. Available at: http://tinyurl.com/huln8aa.

APHIS (2014). JR simplot company petition (14-093-01p) for determination of non-regulated status for Innate™ potatoes with late blight resistance, low acrylamide potential, reduced black spot and lowered reducing sugars: russet burbank event W8. Available at: http://tinyurl.com/zxyb3pt.

Ariño, A., Juan, T., Estopañan, G. and González-Cabo, J.F. (2007). Natural occurrence of *Fusarium* species, fumonisin production by toxigenic strains, and concentrations of fumonisins B1 and B2 in conventional and organic maize grown in Spain. Journal of Food Protection 1: 151-156.

ASAJA (2015). La cosecha del algodón sufre una disminución de la producción debido a su menor peso y ataque de plagas. Available at: http://tinyurl.com/hy6fs8c.

Bakan, B., Melcion, D., Richard-Molard, D. and Cahanier, B. (2002). Fungal growth and *Fusarium* mycotoxin content in isogenic traditional maize and genetically modified maize grown in France and Spain. Journal of Agricultural and Food Chemistry 50: 728-731.

BASF (2016). Sistema clearfield – La tecnología del futuro en el control de malas hierbas. Available at: http://tinyurl.com/hjshomo.

Baulcombe, D., Dunwell, J., Jones, J., Pickett, J. and Puigdomenech, P. (2014). Gm science update – A report to the council for science and technology. Available at: http://tinyurl.com/zm4zqr4.

Brookes, G. and Barfoot, P. (2016). GM crops: global socio-economic and environmental impacts 1996-2014. Available at: http://tinyurl.com/z96rnx4.

Brookes, G. (2014). Weed control changes and genetically modified herbicide tolerant crops in the United States 1996-2012. GM Crops and Food 5: 321-332.

Bucchini, L. and Goldman, L.R. (2002). Starlink corn: a risk analysis. Environmental Health Perspectives 110: 5-13.

Canadian Center for Occupational Health and Safety (2016). OSH answers fact sheets. Available at: http://tinyurl.com/grabnhf.

COCERAL, FEDIOL and FEFAC (2016). Joint letter to EC president Mr. Jean-Claude Juncker. Available at: http://tinyurl.com/glnv7un.

CIBUS (2016). SU Canola and other products. Available at: http://tinyurl.com/gks6vb7.

Cubero, J.I. (2013). Introducción a la mejora genética vegetal, 3rd edition. Mundi-Prensa, Madrid, Spain.

Da Silva, J.G. (2014). FAO calls for a 'paradigm shift' towards sustainable agriculture and family farming. Available at: http://tinyurl.com/kf6f2kc.

Doudna, J.A. and Charpentier, E. (2014). The new frontier of genome engineering with CIRSPR-Cas9. Science 346: 1077.

Duke, S.O. and Powles, S.B. (2009). Glyphosate resistant crops and weeds: now and in the future. AgBioForum 15(3): 231-241.

European Food Safety Authority (EFSA) (2005). EFSA follows up on Bt 10 maize. 2005. Available at: http://tinyurl.com/z9qxgf3.

European Food Safety Authority (EFSA) (2012). Scientific opinion addressing the safety assessment of plants developed through cisgenesis and intragenesis. EFSA Journal 10: 2561.

European Food Safety Authority (EFSA) (2014). Evaluation of the increase of risk of public health related to a possible temporary derogation from the maximum level of deoxynivalenol, zearalenone and fumonisins for maize and maize products. Available at: http://tinyurl.com/jfd483p.

EuropaBio (2014). Approvals of GM crops in the EU – January 2014 update. Available at: http://tinyurl.com/h64b3lp

European Food Safety Authority (EFSA) (2015). Conclusion on the peer review of the pesticide risk assessment of the active substance glyphosate. EFSA Journal 13(11): 4302.

European Commission (EC) (2007). Commission Regulation (EC) No. 1126/2007 of 27 September 2007 amending Regulation (EC) No. 1881/2006 setting maximum levels for certain contaminants in foodstuffs as regads *Fusarium* toxins in maize and maize products. Available at: http://tinyurl.com/zz7l8uh.

European Commission (EC) (2011). Commission Regulation (EU) No. 619/2011 of 24 June 2011 laying down the methods of sampling and analysis for the official control of feed as regards presence of genetically modified material for which an authorisation procedure is pending or the authorisation of which has expired. Available at: http://tinyurl.com/gtbu3zc.

European Commission (EC) (2014). A decade of EU-funded GMO research. Directorate general for research and innovation – Biotechnology, agriculture and food. Available at: http://tinyurl.com/z2qsvug.

European Commission (EC) (2016a). EU register of authorised GMOs. Available at: http://tinyurl.com/3xnedx.

European Commission (2016c). Genetically modified organisms. Europe. Technical Zero. Available at: http://tinyurl.com/j63o36q.

European Council (2003). Regulation (EC) No. 1829/2003 and 1830/2003 of 22 September 2003 on genetically modified food and feed. Available at: http://tinyurl.com/jm9b43j.

European Council (2007). Council Regulation (EC) No. 834/2007 of 28 June 2007 on organic production and labeling of organic products and repealing Regulation (EEC) No. 2092/91. Brussels, Belgium.

European Ombudsman (2016). Decision of the European Ombudsman closing the inquiry into complaint 1582/2014/PHP on the European Commission's handling of authorisation applications for genetically modified food and feed. Available at: http://tinyurl.com/jr4ccb2.

European Parliament (2001). Directive 2001/18/EC of the European Parliament and of the Council of 12 March 2001 on the deliberate release into the environment of genetically modified organisms. Available at: http://tinyurl.com/nff4cqn.

Eurostat (2016). Data explorer. Available at: http://tinyurl.com/jkngz9e.

Estruch, J.J., Warren, G.W., Mullins, M.A., Nye, G.J., Craig, J.A. and Koziel, M.G. (1996). Vip3A, a novel *Bacillus thuringiensis* vegetative insecticidal protein with a wide spectrum of activities against lepidopteran insects. PNAS 63: 5389-5394.

Etzel, M. (2002). Mycotoxins. JAMA 287: 425-427.

Fernandez-Cornejo, J., Hallahan, C., Nehring, R. and Wechsler, S. (2012). Conservation tillage, herbicide use, and genetically engineered crops in the United States: the case of soybeans. AgBioForum 15(3): 231-241.

Fernández-Cornejo, J., Wechsler, S., Livingston, S. and Mitchell, L. (2014). Genetically engineered crops in the United States – USDA (ERR 162). Available at: http://tinyurl.com/yalvql8q.

Food and Agriculture Organisation and International Atomic Energy Agency (FAO-IAEA) (2011). Plant mutation-breeding and biotechnology. Available at: http://tinyurl.com/ztvm3e2.

Food and Agriculture Organisation (FAO) (2014). Technical consultation on low levels of genetically modified (GM) crops in international food and feed trade. Available at: http://tinyurl.com/zfsjcn4.

Food and Agriculture Organisation and International Atomic Energy Agency (FAO-IAEA) (2016a). Joint program – Mission. Available at: http://tinyurl.com/zzbjlq6.

Food and Agriculture Organisation and International Atomic Energy Agency (FAO-IAEA) (2016b). The joint FAO-IAEA mutant variety database. Available at: https://mvd.iaea.org.

Food and Agriculture Organisation/World Health Organisation (FAO/WHO) (2016). Joint meeting on pesticide residues. Summary report. Available at: http://tinyurl.com/hr88tge.

Futtermittel (2016). Europa ist von Sojaimporten abhängig. Available at: http://tinyurl.com/h7ywaf5.

Gil-Humanes, J., Piston, G., Altamirano-Fortoul, R., Real, A., Comino, I., Sousa, C., Rosell, C.M. and Barro, F. (2014). Reduced-gliadin wheat bread: an alternative to the gluten-free diet for consumers suffering gluten-related pathologies. PLoS ONE 9(3): e90898.

GMO Compass (2016). Database. Previously available at: http://www.gmo-compass.org/eng/gmo/db [link obsolete].

Guyton, K.Z., Loomis, D., Grosse, Y., El Ghissassi, F., Benbrahim-Tallaa, L., Guha, N., Scoccianti, C., Mattock, H. and Straif, K. (2015). Carcinogenicity of tetrachlorvinphos, parathion, malathion, diazinon, and glyphosate. Lancet 16: 490-491.

Health Canada (2007). Clearfield bread wheat variety BW7. Available at: http://tinyurl.com/z9bbam6.

Henseler, M., Piot-Lepetit, I., Ferrari, E., Gonzalez-Mellado, A., Banse, M., Grethe, H., Parisi, C. and Hélaine, S. (2013). On the asynchronous approvals of GM crops: potential market impacts of a trade disruption of EU soy imports. Food Policy 41: 166-176.

International Agency for Research on Cancer (IARC) (2016). Agents classified by IARC Monographs Volumes 1-117. Available at: http://tinyurl.com/zsazp7d.

International Service for the Acquisition of Agri-biotech Applications (ISAAA) (2016). GM approval database. Available at: http://tinyurl.com/hhzocle.

James, C. (2006). Global status of commercialized Biotech/GM crops: 2006. Brief 35. Available at: http://tinyurl.com/zt3lmqw.

James, C. (2015). Global status of commercialized Biotech/GM crops: 2015. Brief 51. Available at: http://tinyurl.com/zlhe8sd.

James, C. (2016). 20th anniversary (1996-2015) of the global commercialization of biotech crops and Biotech Crop Highlights in 2015. Brief 51. Available at: http://tinyurl.com/zlhe8sd.

Jones, H.D. (2015). Regulatory uncertainty over genome editing. Nature Plants 1: 1-3.

Kalaitzandonakes, N., Kaufman, J. and Miller, D. (2014). Potential economic impacts of zero thresholds for unapproved GMOs: the EU case. Food Policy 45: 146-157.

Klümper, W. and Qaim, M. (2014). A meta-analysis of the impacts of genetically modified crops. PLoS ONE 9(11): e111629.

Laureates (2016). Letter supporting precision agriculture (GMOs). Available at: http://tinyurl.com/jhdsolp.

Laursen, L. (2016). Will Europe toast GM wheat for gluten sufferers? Nature Biotechnology 34: 369-371.

Mangalassery, S., Sjögersten, S., Sparkes, D.L., Sturrock, C.J., Craigon, J. and Mooney, S.J. (2014). To what extent can zero tillage lead to a reduction in greenhouse gas emissions from temperate soils? Scientific Reports 4: 4586.

Naqvi, S., Zhu, C., Farre, G., Ramassar, K., Bassie, L., Breitenbach, J., Pérez-Conesa, D., Ros, G., Sandmann, G., Capell, T. and Christou, P. (2009). Transgenic multivitamin corn through biofortification of endosperm with three vitamins representing three distinct metabolic pathways. PNAS 106: 7762-7767.

National Research Council (2004). Safety of genetically engineered foods: approaches to assessing unintended health effects. National Academies Press, Washington, DC, USA.

National Academy of Sciences (2010). Impact of genetically engineered crops on farm sustainability in the United States. National Academies Press, Washington, DC, USA.

National Academies of Sciences, Engineering and Medicine (2016). Genetically engineered crops: experiences and prospects. National Academies Press, Washington, DC, USA.

Nicolia, A., Manzo, A., Veronese, F. and Rosellini, D. (2013). An overview of the last 10 years of genetically engineered crop safety research. Critical Reviews in Biotechnology 34(1): 77-88.

Niederhuber, M. (2015). Insecticidal plants – The tech and safety of Bt crops. Available at: http://tinyurl.com/hdbsa8m.

Novozymes (2012). Enzyme cuts acrylamide levels in French fries. Available at: http://tinyurl.com/jeanfcq.

Park, J.R., McFarlane, I., Phipps, R.H. and Cedia, G. (2011). Review: the role of transgenic crops in sustainable development. Plant Biotechnology Journal 9: 2-21.

Pioneer (2016). Corn and soybean herbicide chart. Available at: http://tinyurl.com/jjlwshf.

Pollack, A. (2002). The New York Times. Spread of gene-altered pharmaceutical corn spurs $ 3 million fine. Available at: http://tinyurl.com/z7s6c79.

Queensland University of Technology (2014). Super bananas – world first human trial. GM bananas. Available at: http://tinyurl.com/hllxaoq.

Raven, P.H., Evert, R.E. and Eichhorn, S.E. (1999). Biology of plants, 6th edition. W.H. Freeman and Co, London, UK.

Riesgo, L., Areal, F.J., Sanvido, O. and Rodriguez-Cerezo, E. (2010). Distances needed to limit cross-fertilization between GM and conventional maize in Europe. Nature Biotechnology 8: 780-782.

Rommens, C.M., Yan, H., Swords, K., Richael, C. and Ye, J. (2008). Low-acrylamide French fries and potato chips. Plant Biotechnology Journal 6: 843-853.

Rymal, K.S., Chambliss, O.L., Bond, M.D. and Smith, D.A. (1984). Squash containing toxic cucurbitacin compounds occurring in California and Alabama. Journal of Food Protection 47: 270-271.

Seligman, P.J., Mathias, C.G., O'Malley, M.A., Beier, R.C., Fehrs, L.J., Serrill, W.S. and Halperin, W.E. (1987). Phytophotodermatitis fromcelery among grocery store workers. Archives of Dermatology 123: 1478-1482.

Simplot Plant Sciences (2016). Available at: http://www.simplotplantsciences.com.

Slater, A., Scott, N. and Fowler, M. (2008). Plant biotechnology: the genetic manipulation of plants, 2nd edition. Oxford University Press, Oxford, UK.

Smartt, J. and Simmonds, N.W. (1995). Evolution of crop plants, 2nd edition. Longman Scientific, Burnt Mill, Essex, UK.

Späth, B. (2016). No excuse. Available at: http://tinyurl.com/hmylp6r.

Taheripour, F., Mahaffey, H. and Tyner, W.E. (2015). Evaluation of economic, land use, and land use emission impacts of substituting non-GM crops for GM crops in the US. Available at: http://tinyurl.com/jq7mb7r.

Tribó, F., Mateos, A. and Ojembarrena, A. (2016). Red de ensayos de Pioneer Hi-Bred, Spain, Sl. Unpublished observations of July 4, 2016.

University of Wisconsin Extension Services (2004). Wisconsin farmers and agri-business call for glyphosate (roundup) stewardship. Available at: http://tinyurl.com/hvsju8z.

Ursin, V.M. (2003). Modification of plant lipids for human health: development of functional land-based omega-3 fatty acids. Journal of Nutrition 133: 4271-4274.

US Food and Drug Administration (2016). Acrylamide questions and answers. Available at: http://tinyurl.com/hjunybz.

US Department of Agriculture (USDA) (2015). GAIN report. Spain. Cotton crop update 2015. Available at: http://tinyurl.com/jmv4nos.

US Department of Agriculture (USDA) (2016). Letter from the USDA-APHIS deputy administrator to the investigator. Available at: http://tinyurl.com/zaxacrm.

Usher, S., Haslam, R.P., Ruiz-Lopez, N., Sayanova, O. and Napier, J.A. (2015). Field trial evaluation of the accumulation of omega-3 long chain polyunsaturated fatty acids in transgenic *Camelina sativa*: making fish oil substitutes in plants. Metabolic Engineering Communications 2: 93-98.

Wu, F. (2006). Mycotoxin reduction in Bt corn: potential economic, health and regulatory impacts. Transgenic Research 15: 277-289.

6. Franken-food or techno-fix? – Ethical issues of GMOs in Europe

M. Kaiser
Centre for the Study of the Sciences and the Humanities, University of Bergen, P.O. Box 7805, 5020 Bergen, Norway; matthias.kaiser@uib.no

Abstract

The issue of genetically modified organisms (GMOs) has been dragging on since the 1990s as one of the most complex questions in the governance of the development of science and technology in Europe. In this context and from an ethical-philosophical perspective, this chapter situates the factors that led to the adoption of Directive (EU) 2015/412, and to assess the merits or demerits of the decisions included in the latter, this work analyses the evolution of the debate over GMOs in the EU and its key elements, past and present. Firstly it addresses how policy makers have tackled concern over the potential risks of GMOs to health and the environment and the implications of the risk assessment model selected. This is followed by reflection upon the place occupied in public perception, especially in scepticism, by what have been described, as will be seen, as "ethical considerations". Finally, and prior to reflection upon the contribution of the new Directive, attention is paid to the approach to the question of the coexistence of different cultivation methods.

Keywords: risk management, uncertainty, public perception of GMOs, (un)naturalness, public acceptance

6.1 Introduction

If there ever has been an issue which substantially has influenced the European Commission's policy for science and technology development in Europe, it must have been the topic of genetically modified organisms (GMOs), in particular in relation to crops and food. This issue turned into a governmental nightmare, and one may ask oneself if those in power not felt tempted to follow the advice of Bertolt Brecht: '*Would it not be easier – in that case for the government – to dissolve the people – and elect another?*' True enough, the same could perhaps be said for many of the European member states which had their national discussions, but the crucial point here is that all forms of top-down government failed, and had to give way for governance with an important element of bottom-up input and constraints. The GMO debate in Europe showed major ambivalences (BEPCA, 1997; May, 1999) and has always been a battle, but it was not only a battle between high-level governmental administration and a rebelling public, it was also a battle between those who claimed to possess the relevant expertise, and those who contested this proclaimed expertise. Philosophically speaking, it was the colliding views on the correct and justified notions of rationality, knowledge and science, which were put to the test. Underlying this issue is the question how to manage deep system uncertainties when coupled to high value stakes and the impossibility to clearly separate the facts from the values. This situation already signals the relevance of the post-normal science approach (Funtowicz and Ravetz, 1993), but I shall come back to this later.

DOI 10.3920/978-90-8686-845-2_6, © Wageningen Academic Publishers 2017

6.2 Risks and uncertainties

GMOs might potentially pose a risk to health and the environment. This much was, as far as we know, never denied by anybody involved in policy making. The question was always: do they actually? The claim was that with good scientific risk assessment we will get a grip on these potential risks and get evidence whether a particular GMO actually has harmful side-effects.

Risk assessments of GMOs is in many ways different from risk assessments of e.g. chemicals. We need to be pro-active, and thus cannot count on a large number of data collected over many years. We need to assess the risks before actions are taken, meaning before a GMO is introduced into the market. This is why we normally proceed step-by-step and case-by-case in the assessment of the GMO developments. However, more important in a moral sense is the fact that what we know may be insignificant in relation to what we do not know. In other words, the uncertainties involved may carry more moral weight than the knowledge that we acquire. Let me list some of the ethically important considerations:

First, one is basically talking about a pro-active assessment, not a post-facto assessment. In other words, one needs to know about these risks *before* the GMO has been allowed for widespread use in markets. Second, pro-active assessments need to be coupled to a-priori controls, i.e. to measures that allow keeping the spreading of the GMO under control in pace with our knowledge about the risks. Third, before the GMO debates, living organisms have not been made subject to risk assessment. Risk assessment (other than the risk studies within insurance companies) came to prominence with chemical substances and agricultural industry, and with the nuclear power plants. The complexities and bio-dynamics of living organisms has until then not been the standard topic of risk assessment. Fourth, the uncertainties that accompany all quantitative risk assessments of sufficiently complex systems have already been pointed out in the scientific literature and were not new (Shrader-Frechette, 1991). What also was pointed out in the relevant literature (Beck, 1992) was that conceptualisations and evaluations of risks were essentially value-infected (Shrader-Frechette, 1991), as was the weighing of risks versus benefits through risk-cost-benefit analyses. Examples of failure and the unreliability of risk assessments were published in the literature already in the early 1990's (Wynne, 1991, 1992).

However, this did not shatter the belief in the solidity of risk assessment of GMOs among authorities and proclaimed scientific experts. One should also be aware that on the arena of international trade, as e.g. regulated in the World Trade Organization (WTO), the scientific risk assessments would play a decisive role concerning the question whether or not a country could limit the marketing of such products. The quest was for 'sound science' to support any claim of unacceptable risks, and with sound science one meant basically the standard procedures of quantitative risk assessments. Here is the text from Annex III of the Cartagena Protocol:

> 3. Risk assessment should be carried out in a scientifically sound and transparent manner, and can take into account expert advice of, and guidelines developed by, relevant international organizations.
> 4. Lack of scientific knowledge or scientific consensus should not necessarily be interpreted as indicating a particular level of risk, an absence of risk, or an acceptable risk.

> 5. Risks associated with living modified organisms or products thereof, namely, processed materials that are of living modified organism origin, containing detectable novel combinations of replicable genetic material obtained through the use of modern biotechnology, should be considered in the context of the risks posed by the non-modified recipients or parental organisms in the likely potential receiving environment.

Given the already existing criticisms of risk assessment, the use of the term 'sound' in this connection makes it, so I would claim, into an oxymoron. The rhetoric behind this choice of words thus becomes obvious.

I have said that living organisms constitute complex systems, difficult to model in their dynamics. A basic characteristic of them is deep system uncertainty (cf. Van der Sluijs *et al.,* 2008). When it comes to GMOs these uncertainties were recognised, though experts would place them differently, according to where their field of expertise was.

> Each scientific discipline hence represents a different frame for perception of environmental uncertainty. Molecular biologists refer to the controlled laboratory practice, plant biologists refer to the history of conventional plant breeding, while ecologists argue that the experiences based on the introduction of novel species into new environments need to be the basis for risk assessment of GMOs. The values and implicit normative assumptions need to be fully articulated in order to interpret to what extent a particular scientific disagreement represents 'lack of knowledge' or 'lack of coherence' in the interpretation of strong uncertainties. (Aslaksen *et al.,* 2013: 116)

Similar things could assumedly be said in relation to potential health effects. As it turned out, introducing benchmarks and endpoints was not a straightforward task in relation to GMOs. What is the comparison to a GM-tomato? The tomato from a greenhouse, the tomato from traditional herbicide and pesticide assisted agriculture, the organic tomato or what? For instance, if one aims at some form of biodiversity, then conventional agricultural products may not be a good comparator since they typically score badly on this. Again, concepts like 'substantial equivalence' turned out loaded with problematic (value) assumptions and were contested.

A very contentious issue was and remains to be the question who provides the evidence for assessing the risks? As it turned out, the process of market authorisation expected the industry to provide the relevant evidence, since one could not expect the public to pay for the research. However, the scientific assessments that were/are provided by industry can hardly be ranked as independent research, and thus there is a serious issue how much one should trust these industry studies. This is further complicated by the fact that the raw data of industry studies are typically not openly accessible, and thus not controllable by third parties. Furthermore, there are only a few independent studies performed by public research bodies like universities. Oftentimes when such studies resulted in critical results of a GMO, the researcher(s) were haunted by a seemingly united front of 'experts' who accused these studies not to be sound science.

Another issue of relevance was that sometimes the science, even though it was presented as sound science, in effect was embarrassingly defective. One typical fallacy that occurred was that some scientists seemed unable to distinguish between providing evidence for no harm versus not being able to give evidence for a harm to occur. Absence of evidence for no harm occurring is not the same as evidence of absence of a harm occurring. A report *Absence of toxicity of* bacillus thuringiensis *pollen to black swallowtails under field conditions* claims in its title there is no harmful effect; while they only found '*no significant weight differences among larvae as a function of distance from the corn field or pollen level*' (Wraight *et al.,* 2000). Thus, they only failed to demonstrate a harmful effect, they have not proven that there is none.

I would claim there is an important ethical issue behind all this. I ask what is the ethical use of uncertain science in a setting where the goal is seemingly to create an impression that everything is under control.

Thus I would highlight as my first ethical issue point that the role of science in the GMO debate was and is stretching the limits of what science can achieve; it is used rhetorically and most of the times misused of both pro- and con-groups, including high level administration and politics. They typically ignore the inherent uncertainty and complexity of the issue. On the other hand, sometimes the uncertainties are used to claim that GMOs are inherently unsafe. Neither course is acceptable. Science will not, and cannot be able to provide any assurances that GMO technology is safe. There is the inherent asymmetry between claims of identified harms and claims of identified safety. The first can be straightforward if there is positive evidence for this, the second is an impossible hurdle. Science can help design safer products or help design safer management system of these products or identify variable barriers to prevent unintentional spreading of the GMO. But no scientific finding can establish that a given GMO is safe. Any claim to this effect is unethical. So is the claim that the uncertainty surrounding GMOs is a sufficient ground to abandon this research.

6.3 Ethics and public perceptions

Was the public really only worried about the risks? Is the criticism of GMOs only based on irrational gut-feelings without reasonable arguments and insights? Is it the imaginary of 'Franken-food' (associative of Mary Shelley's Frankenstein), which causes such widespread scepticism? One of the common claims among scientists was that the public (and that often meant only those who did not like GMOs!) was uninformed and built up fear out of lack of knowledge, an idea which was termed the 'deficit model'. However, the time sequence of the Eurobarometers proved differently: over time people acquired more knowledge, but this did not effectively alter the pro- and con-split in the population, it just reduced the segment of those who were undecided, while increasing both the pros and the cons (Hviid Nielsen, 1997). Recently the picture may have become a little less dramatic (Gaskell *et al.,* 2010), but that could perhaps be attributed to the fact that the European Union had not given a green light for the GM-food imports or GMO productions, except in a very few cases (maize in Spain) or restricted usage (animal feed).

What emerged was that the critical attitude towards GMOs was less influenced by risk arguments, then by what people termed 'ethical' considerations (Lassen *et al.,* 2002). People simply asked if there were sufficient benefits (cf. NENT, 1996), and if yes, who would accrue these benefits? And if there were risks – there must obviously be some risks? – who would have to shoulder the risks? And who would pay the costs? How about the externalities, would that be paid by nature again? And is all this talk about substantial equivalence just a trick to smuggle products onto the supermarket shelves which a significant proportion of buyers would not buy if it was labelled as something different, namely as GMO?

One sees immediately that people target ethical principles with these questions, principles having to do with social justice, with autonomy and freedom of self-determination, with respect for nature, with solidarity, and about the trade-offs between risks, costs and benefits. This is in itself quite remarkable since it is not usually the case that the introduction of new technology raises explicitly ethical issues. The nuclear power debate was dominated by risk issues rather than ethics issues, at least until after Chernobyl. Yet there are a few historic examples, as e.g. the introduction of the public telephone, which apparently raised serious moral doubts about the private entering the public space (De Sola Pool, 1983). In principle, what is surprising is not that ethical issues are related to new technologies, as all technology development is based on value choices (Bijker, 1997). What is surprising is rather that ethical concerns typically seem to come as the last resort of techno-scepticism, rather than the first consideration to bring into discussion. People seem even to be a bit reluctant to raise these ethical concerns publicly, perhaps because many feel that this is too subjective a field, a field of soft emotions rather than hard rational arguments[32].

Be that as it may, for me it raises an important second ethical issue in this debate. Given that ethical issues play into the evaluation of GMOs (or any new technology), and given that the publics often lack the articulation and the skills to address these issues upfront, the regulators would have to face the responsibility to provide the fora where ethical issues are explicitly encouraged and centre stage. In other words, to the extent that GMOs trouble people and the state needs to assess the wisdom of introducing these products into the market, it will become necessary to designate platforms where ethical debate is explicitly encouraged and fostered. This would not only include standard stakeholders, but assumedly civic society at large. Here we see that the EU in its Directive 2001/18/EC on the deliberate release of GMOs into the environment went only half way:

> ***Article 29: Consultation of Committee(s) on Ethics***
>
> 1. Without prejudice to the competence of Member States, as regards ethical issues, the Commission shall, on its own initiative or at the request of the European Parliament or the Council, consult any committee it has created with a view to obtaining its advice on the ethical implications of biotechnology, such as the European Group on Ethics in Science and New Technologies, on ethical issues of a general nature.
>
> This consultation may also take place at the request of a Member State.

Ethics was still optional in the regulatory procedures, and ethical advice is in no definite way constraining the final decisions of the Commission.

[32] This is, of course, an ironic use of words, just to make that clear. Often the ethical issue can be a much harder fact – like e.g. unjust distributions of risk – than the perceived scientific risks, which, as we have argued, are systematically ignorant about outrageous events, like – well – black swans.

It is instructive to compare this Directive to the Norwegian Gene Technology Act of 1993. Norway had included both sustainability and societal benefit as additional criteria to risk assessment for allowing the use of GMOs in Norway. A part of this was also ethics, even though the authorities avoided a final clarification of what kind of ethical consideration one would consider relevant.

What I see in these regulatory efforts is a first attempt to be responsive to public worries based on ethical considerations. However, due to a lack of clarity what a justified ethical concern could consist of, and how to go about unveiling these ethical concerns, regulations remained non-binding and detached from wider European publics. One can easily get the impression that the high-level debate on GMOs (of The European Group on Ethics in Science and New Technologies or other bodies) was nothing but window-dressing, while the real weighty arguments on the political level were economic. This I see as paradox: While the public concern was/is to a large extent tied to ethical issues of GMOs, the authorities never developed a policy how to deal with ethics in this area.

6.4 Ethics on a limb?

What was interesting to note was what kind of ethical concerns entered the public sphere. This was not unconnected to the issue of who voiced these concerns, like for instance views coming from religious groups. As soon as ethics was on the agenda, the authorities felt apparently more comfortable with having *Men of Religion* (and this typically implied men, often excluding women) on the board, rather than secular philosophers or other specialists from normative disciplines. From (some of) these corners, intrinsic arguments were voiced against GMOs.

Ethical concerns about genetic engineering may be divided into two basic types. For a *variety of reasons one may think that engineering a life form is morally problematic in* itself, i.e. due to its mode of production or to its source of genetic material. These are then *intrinsic* concerns. But genetic engineering may also be thought of as morally problematic because of its consequences. In that case one talks of *extrinsic* concerns. This can be important since, for instance, risk considerations have no weight, if fundamental intrinsic worries are brought to the fore.

Comstock (2002) argues that all variants of intrinsic arguments against biotechnology could be summarised in the following claim: *'it is unnatural to genetically engineer plants, animals and foods'.*

There is variety of ways how such an argument could be spelled out in more detail. The three most salient ones are:

1. We should not be 'playing God'.
2. Genetic modifications of organisms break down natural species boundaries.
3. Animal biotechnology implies the commodification of all life forms.

Many ethicists have criticised these claims, and indeed, it is not clear how to argue for them. One could, for instance, try to clarify what the ethically relevant difference is between genetic information obtained from a distant species and genetic information that occurs through natural mutation. Both are nothing but a certain sequence of adenin (A), thymin (T), cytosin (C), and guanin (G) in the DNA molecule. The mutant genetic information is, when expressed in an individual, an important tool for selective breeding. In principle, the results of recombinant DNA-techniques could equally well be the results of mutation, given sufficient time. After all, according to evolutionary theory all species have developed from a common core. The dimension of time would become a morally relevant feature for ethical considerations. While time may indeed play a crucial role in regard to the risks (i.e. extrinsic viewpoint) involved in genetic modification (e.g. concerning the 'disrupting' effects of the spread of a genetic modification to the environment), it is difficult to conceive of time as an intrinsic ethical consideration. Therefore this line of defence for a criticism of GMOs based on intrinsic considerations is difficult to maintain.

Intrinsic ethical concerns of a similar kind may also be raised without the reliance on an explicit religious foundation. Respondents in surveys about biotechnology (e.g. in the Eurobarometer, cf. AEBC, 2002; Gaskell *et al.,* 2010) often express concern that genetic engineering is unnatural. The argument then is that the genetic modification of species breaks some natural order between species that is in a finely tuned state of equilibrium. Nature and all that is natural is then assumedly valuable and good in itself, while the tempering with nature, as all forms of genetic engineering where species boundaries are crossed, appears as unnatural and interfering with Nature, and is thus conceived as intrinsically wrong.

Any such argument faces two difficulties that need to be examined. First, what is meant by being natural or unnatural, and second what is good about being natural viz. bad about being unnatural?

It seems quite clear that large part of our food production represent some form of unnatural element in Nature. All domestic species and most agricultural crops are the results of selective breeding. Most of the 'nature' surrounding our cities is cultivated and thus far from being natural. Thus a notion of nature in a general sense seems at least problematic when used in this argument.

More specifically one may hold that it is the genetic boundaries that exist between species that constitute what is 'naturally' given and that need to be respected. The insertion of unrelated genes may change the entire direction of evolution and represent human goals rather than a natural development. Against this view one may hold that species are never static; their genetic outfit changes over time. It can also be claimed that species are not genetically isolated. Certain viruses, for example, carry genetic material from one species to another. Also the notion of species may not be such a firm notion from a biological point of view as many seem to assume; there exist alternative conceptions of what a species is. In sum, the reference to nature and the natural is not such a firm foundation as the argument seems to presuppose. It merely points to the essential social construction of the term 'nature'.

Even if one should accept some notion of natural, one needs to ask why everything natural should be seen as good and everything unnatural as something wrong. Conceptions of nature vary admittedly significantly, from nature as benign and beautiful to nature as threatening, chaotic and pitiless. Charles Darwin for one viewed the works of nature as '*clumsy, wasteful, blundering, low and horribly cruel*'. Nature carries with it a number of characteristics and not of all them need to be unequivocally good.

Furthermore, philosophers since David Hume have recognised what is normally called the 'naturalistic fallacy'. It states that one cannot infer an 'ought' from an 'is', i.e. you cannot deduce a normative statement out of purely descriptive premises. Therefore, a given notion of nature does not carry with it any normative force without further normative assumptions of some kind. It remains unclear what such assumptions could be. Therefore, it seems justified to conclude that arguments about the unnaturalness of crossing species borders in genetic manipulation do not appear to have much ethical weight. They rest on unclear and imprecise language.

It seems there is no really convincing intrinsic argument against animal biotechnology. Obviously the proposed intrinsic arguments point in the direction of some important ethical considerations, but in the end they fail to establish their case. Yet there are two important provisos here. First, the strong initial intuitive appeal of some of these arguments should count as sufficient reason not to disregard them out of hand, but to re-evaluate their relevance in every qualitatively new case. Perhaps an unrestricted and extensive use of genetic engineering at some point indeed crosses borders that are unprecedented and let us forget the basic respect for life that is essential to our moral nature. Second, the mere fact that some of these intrinsic arguments are apparently strongly felt by many people, i.e. represent what sometimes is called their gut-feelings, should count as sufficient reason to not force upon them products that they find morally unacceptable. Even if we are unable to detect sufficient rationality in these reasons to accept them for ourselves, we should out of respect for other humans – humans that may operate with a different sense of rationality – provide them with choices that make a moral life for them possible. This would mean two things: there should be alternatives to food from animal biotechnology on the market, and there should be clear labelling of products from gm biotechnology.

This leaves us to consider the extrinsic arguments against genetic engineering of food products, i.e. arguments that relate to the consequences rather than the nature of our technology. One such line of argument is environmental considerations.

6.5 Uncontrollability and responsibility

From the point of view of the regulator, what is intended is to create a space for the cultivation of GMOs without thereby intentionally diminishing the production space for those who do not cultivate GMOs, and without thereby diminishing regional biodiversity. One wants more variety rather than less.

> From 1999 onwards, policy-makers started to continuously revise the legal conditions under which GM crops and agro-food products were allowed to be used in the EU to slow down further erosion of public and market confidence... The precautionary principle, post-market environmental monitoring and traceability were legally adopted as ways to cope with scientific uncertainties. New institutions such as the European Food Safety Authority (EFSA) were created to provide independent, objective and transparent science-based advice on the safety of agro-food biotechnology applications. Labelling and traceability of GM products became mandatory to ensure consumers' freedom of choice. Because the maintenance of different agricultural production systems is a prerequisite for providing a high degree of consumers' choice, a coexistence policy was adopted in the EU (Devos *et al.*, 2008).

Coexistence of traditional agriculture with the GMO planting was intended and rules for this to be worked out. What one wants to avoid is adventitious mixing of non-GM with GM crops. In principle the coexistence of different variations of a species should be possible, as one has relatively reliable information on pollination distances. As it turned out, different countries were operating with different safety distances and different measures. And even if people tried to focus on scientific 'facts' underlying it all is a basic value conflict about what agriculture one wants.

> The techno-scientific discussion about isolation distances is in fact hiding an underlying discussion about the type of agriculture wanted in the EU. Thereby, it is debated whether GM crops might play a role in the type of agriculture wanted and whether they might contribute to the construction of a sustainable system of crop production. On an even more fundamental level, one can detect a conflict of values pertaining to the importance of individual freedom of choice and to the trust in markets as regulators of consumer preferences. It may be argued that the broad range of isolation distances, which is supposed to satisfy standards of EU legislation, is in fact reflecting a coexistence paradox that effectively accommodates an irreconcilable divergence of positions towards GM crops' (Devos *et al.,* 2008).

Another ethical issue underlying this debate is the question of responsibility. One may apply the newcomer principle and make the GMO-crop-adopter responsible for all possible measures to prevent adventitious mixing that occurs as a consequence. We know by now that this mixing occurs even at distances hitherto believed to be safe, and traces of the GMOs are found in other varieties (see e.g.: http://tinyurl.com/n3n3brl). Ascribing this sort of forward-looking responsibility implies a sophisticated ethical position and epistemic analysis of social experiments (Robaey, 2016). It also implies a strong version of the Precautionary Principle, e.g. in the form of the UNESCO definition: *'When human activities may lead to morally unacceptable harm that is scientifically plausible but uncertain, actions shall be taken to avoid or diminish that harm'* (UNESCO/COMEST, 2005).

It is obvious that this is a position, which is not universally accepted, in fact strongly opposed by some states, as e.g. the USA, even though it has a backing in European treaties as e.g. the Maastricht and especially the Nice Treaty.

6.6 Directive (EU) 2015/412

What seemingly happened was that introducing GMOs to European agriculture turned more and more into a very complex affair without reducing public opposition at all. Redress to pure 'objectivity' was not possible given all the different and colliding positions, even from within the sciences, not to mention the ethical objections. The usual objective would have been to converge towards a common European policy all across the European member states. But the publics in the different member states reacted quite differently, and no state was willing to risk a deep conflict with its own people over an issue which does not seem to be that crucial for national economies. In fact, organic agricultural products were getting higher prices, and the market niches for these were expanding. The division between, say Spain on the one hand (growing GM maize) and Austria on the other (refusing any GMOs in its country), was all too obvious. Given this background, responding with allowing different strategies to be pursued by the different member states seems both politically wise and strategically sensible.

This was done in Directive 2015/412. What this Directive now opens for is for each member state to find their own policy on GMOs so that it becomes possible to accommodate even extreme positions – like Spain and Austria – without giving up the assessments provided on the European basis. The Directive is a compromise of 'Realpolitik'. It is perhaps – to speak tongue in cheek – short of Brecht's wish to dissolve the European people and elect another.

There are still aspects which I find disappointing, given the long history of the GMO debate, and the insights collected along the way. First, it is clear that the EU is not withdrawing anything about the proclaimed objectivity of the risk assessments of GMOs. It is not re-defining the task of the EFSA in the light of the impossibility to separate out the values from the facts in this kind of risk assessments. It may be because of the international context like the WTO or the Cartagena Protocol that the EU finds it impossible to give up its appeal to sound science and scientific objectivity. Or it may be because of a deep seated ideology about science and technology that makes it impossible to loosen these claims. Yet, what I would have liked to see here is some admission that questions of this kind cannot be decided on objective scientific basis alone, but that the management of uncertainty and the fact-value entanglement makes it necessary to address issues, which are not based on the expertise of a small elite. The limits of the belief in scientific objectivity are not addressed but rather maintained.

Second, connected to the first point I find it disappointing that the EU has refrained from explicitly stating a role for ethics in democratic decision processes. There is no proper platform mentioned where ethical issues should be discussed. It is apparently acknowledged that member states may have different positions here, but within the European decision making ethical issues seem to fall out. It would, I believe, have been a chance to strengthen the role of the European Parliament, and to delineate the format of good public consultations. Neither has happened, and ethics remains a somewhat nebulous concept in European politics.

Third, on the basis of the above two points I would have expected a bolder move towards a recognition that quality assessments in such cases should rest on extended peer reviews, rather than being restricted to a few experts selected on questionable criteria of relevance in disciplinary background. This would amount to recognise that the kind of advice that European S&T politics needs is what is commonly called the post-normal science approach, introduced by Silvio Funtowicz and Jerome Ravetz in the early 1990's (cf. Funtowicz and Ravetz, 1990, 1993, 1999). When value stakes are high, system uncertainties are also high, decisions are urgent, and facts and values are controversial in society, the post-normal science approach is the relevant outlook and argues for extended peer reviews as a crucial element of quality assurance. In post-normal science, knowledge claims are always contextualised. I would like to see more analyses and recognition what strong contextualisation can achieve in debates about science and technology in Europe. With strong contextualisation new ethical issues and dilemmas arise as a result of the growing power of society. It builds on a bottom up perspective in contrast to a top-down approach, and under certain circumstances results not in answers of right or wrong, but in a need of open and transparent models for communication, critical reflection and further development.

The science advisor to the Prime Minister of New Zealand, Sir Peter Gluckman, has expressed this insight and given it some publicity in the scientific press:

> I have come to understand that the primary functions and greatest challenges for a science adviser are providing advice not on straightforward scientific matters, but instead on issues that have the hallmarks of what has been called post-normal science (Funtowicz and Ravetz, 1993). These issues are urgent and of high public and political concern, the people involved hold strong positions based on their values, and the science is complex, incomplete and uncertain. Diverse meanings and understandings of risks and trade-offs dominate (Gluckman, 2014: 163).

The procedural ethics step following from this insight would be to design platforms and tools for extended peer reviews, moving the issue from the shoulders of individual advisors to the broader platform of a societal process. It is this step, which I would have liked to see in the new Directive. Short of that, it is clear that the European debate on GMOs, the span in narratives from Franken-food to Techno-fix, provides unprecedented important lessons on the role and inevitability of ethics in matters of science and technology. I am confident that ethical issues will remain centre stage in the debate of GMOs in Europe.

References

Agriculture and Environment Biotechnology Commission (AEBC). 2002. Animals and biotechnology: a report by the AEBC. Department of Trade and Industry, London, UK.

Aslaksen, I., Glomsrød, S. and Myhr, A.I. (2013) Post-normal science and ecological economics: strategies for precautionary approaches and sustainable development. International Journal Sustainable Development: 16: 107-126.

Beck, U. (1992). Risk society: towards a new modernity, Vol. 17. Sage, Thousand Oaks, CA, USA.

Bijker, W.E. (1997). Of bicycles, bakelites, and bulbs: toward a theory of sociotechnical change. MIT press, Cambridge, MA, USA.

Biotechnology and the European Public Concerted Action group (BEPCA) (1997). Europe ambivalent on biotechnology. Nature 387: 845-847.

Comstock, G. (2002). Ethics and genetically modified foods. In: Ruse, M. and Castle, D. (eds.) Genetically modified foods. Debating biotechnology. Prometheus Books, New York, NY, USA.

De Sola Pool, I. (1983). Forecasting the telephone: a retrospective technology assessment. Ablex Publishing, New York, NY, USA.

Den Nasjonale Forskningsetiske Komité for Naturvitenskap og Teknologi (NENT) (1996). Kvikklaks og teknoburger. Sluttrapport fra lekfolkskonferansen om genmodifisert mat. NENT, Oslo, Norway.

Devos, Y., Demont, M., Dillen, K., Reheul, D., Kaiser, M. and Sanvido, O. (2008). Coexistence of genetically modified (GM) and non-GM crops in the European Union. A review. Agronomy for Sustainable Development 29(1): 11-30.

Funtowicz, S. and Ravetz, J. (1990). Uncertainty and quality in science for policy. Kluwer Academic, Dordrecht, the Netherlands, 244 pp.

Funtowicz, S. and Ravetz, J. (1993). Science for the post-normal age. Futures 26(6): 568-582.

Funtowicz, S. and Ravetz, J. (1999). Post-normal science – An insight now maturing. Futures 31(7): 641-646.

Gaskell, G., Stares, S., Allansdottir, A., Allum, N., Castro, P., Esmer, Y., Fischler, C., Jackson, J., Kronberger, N., Hampel, J., Mejlgaard, N., Quintanilha, A., Rammer, A., Revuelta, G., Stoneman, P., Torgersen, H. and Wagner, W. (2010). Europeans and biotechnology in 2010 – Winds of change? A report to the European Commission's Directorate-General for Research. Available at: http://tinyurl.com/hhkzhcz.

Gluckman, P. (2014). The art of science advice to government. Nature 507: 163-165.

Hviid Nielsen, T. (1997). Behind the color code of 'no'. Nature Biotechnology 15: 1320-1321.

Lassen, J., Madsen, K.H. and Sandøe, P. (2002). Ethics and genetic engineering – lessons to be learned from GM foods. Bioprocess and Biosystems Engineering 24(5): 263-271.

May, R. (1999). Genetically modified foods: faults, worries, policies and public confidence. In: Note by the UK Chief Scientific Adviser. Office of Science and Technology, London, UK.

Robaey, Z. (2016). Gone with the wind: conceiving of moral responsibility in the case of GMO contamination. Science and Engineering Ethics 22: 889.

Shrader-Frechette, K. (1991). Risk and rationality. University of California Press, Los Angeles, CA, USA.

UNESCO/COMEST (2005). The precautionary principle. A report of an *ad hoc* working group. UNESCO, Paris, France. Available at: http://tinyurl.com/k7883v9.

Van der Sluijs, J.P., Petersen, A.C., Janssen, P.H., Risbey, J.S. and Ravetz, J.R. (2008). Exploring the quality of evidence for complex and contested policy decisions. Environmental Research Letters 3(2): 024008.

Wraight, C.L., Zangerl, A.R., Carroll, M.J. and Berenbaum, M.B. (2000). Absence of toxicity of *Bacillus thuringiensis* pollen to black swallowtails under field conditions. Proceedings of the National Academy of Sciences of the United States of America 97: 7700-7703.

Wynne, B. (1991). Knowledges in context. Science, Technology, and Human Values 16(1): 111-121.

Wynne, B. (1992). Uncertainty and environmental learning: reconceiving science and policy in the preventive paradigm. Global Environmental Change 2(2): 111-127.

7. Genome editing: does it move the goalposts on the GM playing field?

D.M. Bruce
Edinethics Ltd., 11/6 Dundonald Street, Edinburgh EH3 6RZ, United Kingdom; info@edinethics.co.uk.

Abstract

In the last five years gene editing has emerged as a powerful scientific tool potentially capable of achieving a range of genetic modifications in plants, animals and recently even human embryos. This paper explores whether or not in addition to opening up the scientific potential of genetic modification (GM), but it might open its ethical and social acceptability to publics, retailers, regulators and policy makers. It draws upon our original ethical findings in the 1990s, subsequent assessments of potential acceptability of novel biotechnologies, and long-term public engagement using the Democs card game. It reviews seven main values underlying the major concerns about GM food and crops. It then looks at the nature of gene editing types, set in the context of the range of ways in which the genetic makeup of plants and animals have been altered hitherto. It considers if the new techniques alter some of the basic ethical objections which some apply to GM crops, especially to transgenesis, and whether this would meet with a greater public acceptance. The regulatory implications are briefly considered, requiring a more sensitive system than is currently applied to GM organisms. The largest factor to acceptance may be the perceived usefulness of the application to humans, which might tip the balance against perceptions of risk and uncertainty. The results of playing a Democs card game on GM crops with MSc students over a period of ten years reveals a cautious view but open to the possibility of modification.

Keywords: genome editing, GM crops, promise, ethics, public acceptance

7.1 Introduction

When GM crops began commercial use in 1996, the supporting rhetoric was its unprecedented precision and the unlimited scope, compared with the uncertainties and restricted range of selective breeding. When examined more closely, however, methods such as random ballistic insertion did not seem like precision. In addition, many of the more appealing applications proved difficult, like enhanced growth rate, nitrogen fixing, stress tolerance and nutritional enhancements. Relatively few second and third generation crops have been commercialised. The vast majority of acreage of GM crops are still two 20-year old agronomic traits, herbicide tolerance and insect resistance, applied to only four crops worldwide. These two main traits are of little tangible benefits to consumers or retailers, and were applied without labelling. In Europe, the lack of consumer choice proved highly controversial, along with strong NGO opposition, and a widespread public concern about possible long term environmental and health risks. The resulting public backlash has meant that almost no genetically modified foods are available in Europe and only one maize crop for animal feed is grown commercially.

DOI 10.3920/978-90-8686-845-2_7, © Wageningen Academic Publishers 2017

Recently, however, genome editing has emerged as a powerful scientific tool which has the potential to deliver belatedly the precision and the wide range applications which GM crops have generally failed to realise. In particular, the CRISPR/Cas method is relatively easy, cheap and is proving notably versatile. It may enable genetic changes in food animals, which had not hitherto been considered worthwhile by the animal breeding industry, and also medical applications such as using sheep or pigs as models of human disease. From an ethical point of view a key difference is that by altering individual nucleotides in the DNA sequence, gene editing does not introduce 'foreign' DNA into the crop. This has prompted hopes in its proponents that genome editing might change the ethical picture of genetically altered organisms (GMOs). Would genome editing now find acceptance with the European public?

7.2 Value criteria for considering genetically modificated crops and food

From 1993-98, an expert working group was set up under the auspices of the Society, Religion and Technology Project of the Church of Scotland, which was the first in depth ethical study in English into non-human genetic modification (Bruce and Bruce, 1998). The Engineering Genesis study, published in 1998, identified most of what became the major issues of concern when the public crisis erupted shortly afterwards in February 1999. A major insight was what has subsequently become known as the 'lightning rod' effect in which GM draws attention a wide range of underlying questions about food, agriculture, risk, public trust and involvement, commercialisation and much else. These were summarised in ten value criteria, representing the key arguments presented for and against genetic modification.

1. Scientific rationality – GM as the next step in technological progress.
2. Commercial values – Economic growth, jobs, competitive advantage.
3. Theological/philosophical – Is switching genes across species right or wrong?
4. Ideological – Should we have industrial or organic agriculture?
5. Risk – When do we know enough over uncertain outcomes? 'Will it harm my child?'
6. Resource/food security – Do we need GM technology 'to feed the world', or not?
7. Development – Can GM deliver justice, nutrition and livelihoods for the global poor?
8. Benefits – Are these primarily for the interests of companies, or citizens and consumers?
9. Trust – Government, regulators, scientists, NGO's; is there any 'honest broker'?
10. Control and participation – Who decides if GM goes ahead, against what criteria?

In summary, how people respond to the GM crops is likely to depend upon which ones amongst these criteria they consider to be the most important. Naturally, this will vary from person to person, and with different social contexts, countries, and organisations. But these provide a check list against which to consider the complexities of genetic modification.

These criteria are applicable to the genome editing of crops and animals. At this early stage of development of the technology, it would be premature to speculate on the impact of genome editing on many of the wider contextual questions as #6 – #10, or on the current status of risk in these techniques (#5). This paper focuses on whether genome editing alters the fundamental ethical questions (#3), the agricultural (#4), and the benefits (#8).

7.3 Is genome editing 'genetic modification'?

Genome editing re-opens the question of exactly what is, and is not, genetic modification? This is a complex question since there are a wide range of genetic interventions, not all are considered as necessarily 'GM'. The presumed base line is the genetic sequences and mutations which exist in nature as 'wild-type', and the use of selective breeding over many generations to emphasise the role of certain traits which the breeder deems to be important. You might call this 'selected naturalness'. This is a form of genetic modification, but seen as natural, even though the products may be hugely different from the originating wild types.

At the other extreme is *transgenesis*, adding a gene from a different species which could not occur naturally. Some make a distinction of a greater degree of 'unnaturalness' if the gene is from a different kingdom, e.g. from fish or bacteria, compared with a gene still within the same kingdom, e.g. a daffodil gene into rice.

In between there are modifications within the same species or closed related ones. One case is *cisgenesis* where genes are artificially transferred between organisms that could otherwise be conventionally bred. Another is *intragenesis* in which new combinations of DNA sequences are created within the species. Genome editing can perform targeted mutations within the organism itself, using enzymes or other non-genetic means to cut and re-insert base pairs in its own DNA. These are generally *cis-* or *intra-genesis*, because they do not involve the introduction of DNA from species with which the organism cannot breed.

At its simplest, genome editing may create a mutation of the species which may exist in the same or a related species, but which the individuals in question do not have. Hornless cattle have been bred by editing into dairy cattle a gene known to be responsible for hornlessness in Aberdeen Angus beef cattle (Recombinetics, 2016). It is possible to achieve the same result by crossing breeds and selecting the hornless animals but would take many years. This might be termed 'selected naturalness'. Genome editing achieves the same result very much faster, in what might be termed 'induced naturalness'. If the process is not seen as ethically objectionable, and if the result is possible to be achieved by selective breeding, it could be argued that there should be little reason to object *on principle* to such examples of genome editing.

A slightly different case is creating a mutation which confers resistance to the disease African Swine Fever into the genome of a typical UK pig. This mutation is found in some African warthogs, but not in domestic pigs (Lillico *et al.,* 2013). The same could not be achieved by normal breeding, because the two species, while close, do not interbreed. Given that both are subject to the disease, it would seem churlish to object on the grounds of being 'a genetically unnatural product'. Perhaps our human classification into kingdom, genus, species etc. is not always be helpful as a reflection of our interaction with nature.

In this case, it is also possible that the mutation introduced by a genome editing process already exists in the domestic species, but has never been observed (personal communication, 2015). We only know the genomes of a very few individuals compared with the whole population, and spontaneous mutations occur all the time. Somewhere 'out there' the mutation might exist in one or more individuals, or may have done in the past, or might do at some time in future. This situation might be termed 'unknown naturalness'.

The ethical objection would be to the *deliberate* creation of a mutation which could exists naturally but is not known to. One might object on grounds of risk that the researcher does not know that it might be a deleterious mutation, but here we are considering the principle of doing so. It would seem strange to object to this ethically, if one accepts the deliberate direction of selective breeding by marker assisted selection, or random mutagenesis.

The most problematic case of genome editing is the introduction of multiple changes, in which the resulting plant or animal would definitely not exist otherwise nor be unachievable by any natural process. An extreme example is the attempt to switch off a large number of pig genes to avoid a violent rejection of a pig's heart if transplanted into the human body (Reardon, 2015). Here the focus is not on the method but the product. Is it ethical to create such a pig, whether by genome editing or any other method?

Genome editing may be compared with the practice of random mutagenesis in which a plant is subjected to irradiation or chemicals to create random mutations within its own genome, and then testing to see if any of the resulting plants have advantages. This has sometimes been successful. The Golden Promise barley variety is one example. The randomness of the method gives it far more uncertainty and more potential to disrupt the genome far more than even by transgenesis, but it is not regarded as genetic modification by the EU regulations, since it does not involve bringing genetic material into the plant from outside. This is significant, because, since genome editing offers *targeted* mutation of the genome, it would be anomalous if genome edited organisms were regarded as a GMOs, if the products of *random* mutagenesis remains outside GMO regulation.

While the picture is complex, there are thus some important ethical differences between genome editing and what may be called 'conventional' GM. It is argued that genome edited plants should not simply be treated in regulations as if they were GM crops, unless a clear majority of EU population perceived genome edited plants to be no different from GM. A more flexible mode of regulation will be needed. This will also need to address who is allowed to perform genome editing. It is a relatively cheap and easy technique which can be done in someone's garage. As a precaution, for the time being we should probably allow only genome editing by licensed researchers, as is the case with animal research.

7.4 Would genome editing find acceptance with publics?

One way of considering these interventions is naturalness. Although questions of natural and unnatural are notoriously difficult in ethics, there are important insights to be had (Bruce and Bruce, 1998; Nuffield Council on Bioethics, 2015). For some, transgenesis poses ethical objections if it is deemed that it is wrong to mix genetic material 'unnaturally', in the sense of a process of genetic mixing that cannot happen in nature. But can it be thought wrong to create 'unnatural' genetic mixes in the sense of a mixture not known to exist in nature, but by a process within the same organism? Is the ethical objection to the unnatural act of transgenesis, or to a genetically unnatural product? Genome editing avoids the first problem but not necessarily the second.

Depending on the level of genetic changes people might be prepared to accept, some examples of genome editing may be acceptable and others more problematic. Here is a gradation of what might be acceptable, starting with minimal acceptability in case A, only selective breeding, through to accepting full transgenesis from any species in case G.

A. Only naturally alteration of a plant's natural genetic make-up.
B. *Artificially* creating a natural mutation or breed-able mutation.
C. Creating a mutation from another species.
D. Creating a mutation not known to exist anywhere.
E. No mixing of genes across species, but OK within species.
F. Mixing genes across fairly close species.
G. Mixing genes across any species.

Acceptability will depend on the extent to which attitudes are hardened into an embedded cultural perception, or remain to some degree 'plastic' in people's minds. For some, it would still be considered 'tampering with genes', even within the genome, and so be unacceptable. It is by no means clear, however, that this perception is the principal view in Europe. Many might be less dogmatic, especially if the product looked useful enough. This might also depend on the nature of the benefit – for example improvements in food animal production, compared with an animal edited for purposes of human medicine. Would people approve of pigs edited to be resistant to a prevalent pig virus, and would they then eat them?

It has been suggested that the acceptance or rejection of novelty and innovation in biotechnology are seen to be dependent on a complex measure of factors, which constitute an unspoken social contract, in the sense that the innovation has to fulfil a set of conditions in order to gain a wide public acceptance (Bruce, 2002). A society is prepared to embrace technology to deliver certain benefits, and will accept a certain degree of risk and adaptation of life styles, provided certain basic conditions are fulfilled. These conditions include:

- Values: does it uphold or challenge basic values?
- Familiarity: is the technology familiar and understood, socially embedded?
- Comparison: if it is unfamiliar, has something like it gone wrong before or proved reliable?
- Control: how much do we feel in control of the technology or the risks?
- Trust: how much do we trust those who are in control?
- Vision: how much do we share their values, motivations and goals?
- Choice: is it a voluntary or imposed risk?
- Frequency of risk: if it's too frequent it's unacceptable.
- Magnitude of risk: there is a much greater aversion to high consequence risks.
- Immediacy of risk: is it noticeable or does it creep up insidiously?
- Benefit: does it offer realistic, tangible benefits to the consumer?
- Profile: has it been given a positive or negative image in the media?

If several of these factors are not fulfilled, the technology is unlikely to be accepted. Genome editing is likely to suffer from a comparison with genetic modification, and many of the concerns over control, trust and familiarity are therefore likely to be mapped over from GM. But the value issues may be less problematic than for GM. Perceived risk and uncertainty may be a significant barrier, but could be less than for GM crops if a good case can be made that off-target events pose no significant long-term risk. At present it is too early to say, but these lie outside the scope of the present paper. Genome editing procedures may be more reliable than GM methods, but subtle risk distinctions may not easily be conveyed to publics.

On benefits, genome editing has the potential to be much more relevant to consumers than GM has been. Finally, at the present time, the media profile is fairly positive and has not been the subject of major negative campaigns in the press. Thus there is some hope that perceptions of genome editing may be less negative than for GMOs. But this illustrates a point that whether publics, regulators or Parliaments find genome editing acceptable or not will depend on many more factors than a detailed exercise in principled moral reasoning.

The largest factor may be in the perceived usefulness of the application to humans, which might tip the balance over questions of risk and uncertainty. In both animals and crops, if the lead applications of genome editing are agronomic traits with no tangible benefit to European consumers, the perception is likely to be that this is just another version of GM. If producers wish to present genome editing as something different which should make a sea change in public acceptance, it will need to look very different in terms of the applications that are offered. It remains to be seen whether the commercial aims of the principal actors will once again reduce the field to a few agronomic traits, or whether this time the more imaginative applications will also attend to factors of more value to consumers, and particularly to address the needs of the global poor and malnourished.

A further issue is the extent to which those with agency to enable genome edited foods and other products to be marketed, notably the principal retail chains, are swayed more by those who object than those who might be willing to try. One might think a genome edited pig resistant to a deadly disease was a good thing, but would one then buy the ham, if this became a normal item on the supermarket shelf?

At the moment these questions are largely untried with general publics. At the 2015 EURSAFE conference we reported the results of playing a Democs card game on GM crops with MSc students over a period of ten years, which revealed a cautious view but open to the possibility of GM crops (Bruce, 2015). As a preliminary exercise, a set of games played in March 2016 included some additional questions on genome editing. Opinions among the students varied as to whether this was less ethically problematic than GM, and whether examples like the resistant pig were acceptable. This initial result will need further games to explore.

7.5 Conclusions

Avoiding transgenesis should address the concerns of those whose basic concern was about mixing genes across species, or violating evolved or God-given 'barriers'. Similarly, the notion that the edited sequence is capable of occurring naturally would be attractive if one's objection was to creating an unnatural gene construct. But if one's objection was to any genetic alteration beyond what is achievable by selective breeding, genome editing might be acceptable only under a very limited set of circumstances. It would not be accepted at all by someone whose primary concern is that of scientists 'tampering with our food'.

Thus genome editing in food crops and animals does represent ethical differences from transgenesis and should not automatically by included within the EU GMO regulatory framework. But is not expected to be a panacea which solves 'the GM problem'. Rather, it is likely to receive a range of responses, from acceptance to continued opposition, depending on what factors matter most to people.

Overall, although genome editing has the potential to be a 'game changer' on biotech interventions in food, from a technical point of view, it will need very careful managing for the goalposts to move very far with publics.

References

Bruce, D. and Bruce, A. (1998). Engineering genesis. The ethics of genetic engineering in non-human species. Earthscan, London, UK.

Bruce, D.M. (2002). A social contract for biotechnology – shared visions for risky technologies? Journal of Agricultural and Environmental Ethics 15: 279-289.

Bruce, D. (2015). Changing attitudes to GM crops? In: Dumitras, D., Jitea, I.M. and Aerts, S. (eds.) Know your food: food ethics and innovation. Wageningen Academic Publishers, Wageningen, the Netherlands, pp. 283-288.

Lillico, S.G., Proudfoot, C., Carlson, D.F., Stverakova, D., Neil, C., King, T.J., Ritchie, W.A., Tan, W., Mileham, A.J., McLaren, D.G., Fahrenkrug, S.C., Whitelaw, C.B.A. (2013). Live pigs produced from genome edited zygotes. Scientific Reports 3: 2847.

Nuffield Council on Bioethics (2015). Ideas about naturalness in public and political debates about science, technology and medicine: analysis paper. Nuffield Council on Bioethics, London, UK.

Reardon, S. (2015). Gene-editing record smashed in pigs. Nature News. Available at: http://tinyurl.com/p5tfu6r.

Recombinetics (2016). Available at: http://www.recombinetics.com/node/91.

8. Genetically modified crops in India: some observations

G. Pakky Reddy
Executive Director, Agri Biotech Foundation, Acharya N.G. Ranga Agriculture University, 500 030 Rajendranagar, India; agribiotechfoundation@gmail.com

Abstract

India is ranked 12th in the world in Biotech and second in Asia. This chapter analyses the successful evolution of Bt cotton, one of the important commercial crops grown in India, and compares it with the difficulties being experienced with other Bt crops, like Bt brinjal. Detailed information is provided regarding the transgenic crops that are under development and field trials in India and there is explanation of the way in which debate over GMOs is developing in India. In relation to the latter there is analysis of, amongst other things, public engagement, the involvement of different constitutional bodies, regulatory mechanisms and the procedure for approval of GMOs.

Keywords: biotechnology regulatory authority of India, GMO approval procedure, Bt brinjal, Bt cotton, GM debate

8.1 Introduction

Genetic Modification (GM) may be defined as altering DNA of an organism. This may involve altering an existing gene or introducing a new gene into an organism. Biologists believe that the GM technology is a powerful tool that has a potential to change the land scape of agriculture worldwide. Although scientists have been working for the last several decades, the GM product in the form of Flavour Saved Tomato was first released commercially in 1996. Since then a number of GM crops entered the market in a number of countries. The area under GM crops worldwide increased phenomenally starting from 1.7 million hectares in 1996 to 181 million hectares in 2014 (ISAAA, 2014).

Indian Government recognised the potential of GM crops and encouraged research in a big way. A number of public institutions and private companies have been engaged in GM technology development. Prominent Indian Institutions engaged in GMO technologies are Indian Council for Agriculture Research, Council for Scientific and Industrial Research, Department of Biotechnology (DBT), Department of Science and Technology (DST) etc. Among these agencies the DBT plays a crucial role in terms of developing policy guidelines and research funding. The research on GMOs received a big fillip after the establishment of DBT in 1986. During the last 30 years the department set up a number of research organisations and funded a number of research projects to individuals and institutions. The National Biotechnology Development Strategy 2015-2020 released by the DBT in December 2015 recognises that biotechnology has the potential to be a globally intellectual enterprise of humankind. The document envisages (Ministry of Science and Technology, 2015):

- *providing* impetus to fulfilment of the potential for a new understanding of life processes and utilising the knowledge and tools to the advantage of humanity;
- *launching* a major, well directed effort backed by significant investment for generation of biotech products, processes and technologies to enhance efficiency, productivity, safety and

DOI 10.3920/978-90-8686-845-2_8, © Wageningen Academic Publishers 2017

cost effectiveness of agriculture, food and nutritional security; affordable health wellness; environmental safety; clean energy and biofuel; and bio-manufacturing;
- *empowering* scientifically and technologically, India's incomparable human resources;
- *creating* a strong infrastructure for research, development and commercialisation for a robust bio economy and;
- *establishing* India as a world class bio-manufacturing hub for developing and developed markets.

8.2 Indian biotech industry at a glance

The Indian Biotech industry holds about 2% share of the global biotech industry. The biotechnology industry in India, comprising about 400 companies has grown three fold in the last five years to reach a revenue level of US$ 4 billion. It is expected to grow from the current level to US$ 100 billion by 2025 by a growth of about 30% according to Association of Biotechnology Led Enterprises. India's Biotech Industry comprises of biopharmaceuticals, bio services, bio agriculture, bio industry and bioinformatics. Biopharma is the largest sector contributing about 62% of the total revenue followed by bioservices (18%) and bioagri (15%). Presently India is ranked 12th in the world in biotech and second in Asia. The high demand for different biotech products has opened up scope for the foreign companies to set up base in India. The Indian government on the other hand aims at scaling up the number of start-ups in biotechnology sector to 1500-2000 in the next two to three years from present 500 units (Indian Brand Equity Foundation, www.ibef.org).

8.3 Bt cotton, the successful story

Cotton is one of the important commercial crops grown in India. It occupies a significant place in Indian agriculture and industrial (textile) economy of the country. Nearly 15 million farmers spread over 10 states are dependent on cotton cultivation for their livelihood. At the end of 2002 total area under cotton crop in the country was 9.10 million ha. which accounted for 6% of net cropped area. India which is the third largest producer of cotton in the world behind China and the USA, accounts for 25% of world acreage but only for 14% of world production. Despite being one of the largest cotton growers in the world, cotton yield in India was one of the lowest in the world because of severe pest ravages and its predominant cultivation under rainfed cultivation. Available estimates show that out of total pesticide consumption of Rs.2800 crores in Indian agriculture, about Rs.1600 crores were spent on cotton alone of which Rs.1100 crores were spent only to control bollworms (Narayanamoorthy and Kalamkar, 2006).

Bt. cotton, the first GM crop in India was released for commercial cultivation in April 2002 to control American bollworm. The *Cry 1 AC* gene obtained from soil bacterium *Bacillus thuringiensis* was introduced into a cotton hybrid. Three hybrids consisting of a single Cry 1 AC gene developed by Mahyco-Monsanto Biotech, a joint venture of Indian and US firms, were released in the first instance. By 2006 a two gene event, commonly known as Bollgard-II developed by Mahyco and sourced from Monsanto was released to offer control against Spodoptera in addition to bollworm complex. By 2014 there were about 1500 Bt. cotton hybrids available in the market released by about 50 companies. Most of these companies sourced the technology from Mahyco-Monsanto Biotech.

The impact of Bt. cotton is visible. The number of farmers cultivating Bt. cotton increased from 50,000 in 2002 to 7.2 million in 2012. The area under Bt. cotton cultivation expanded quite rapidly from 24,000 hectares in 2002 to 11.60 million hectares in 2014 accounting for 91% total cotton area in the country. This is the fastest rate of adoption of any new agri technology in the history of Indian agriculture. Bt. cotton contributed positively in a number of ways. The productivity level increased from a meagre 190 kg (one of the lowest in the world) to 537 kg per hectare. Total cotton production has gone up to 4.0 million bales per annum from 1.36 million bales in 2002. This is the highest in the world. The consumption of insecticide is reduced by 40%. Net income levels of farmers have gone up. Thus, the overall rural economy has improved. A sample survey conducted by Ajay Thutupalli revealed that 60% of the respondents said that their profits increased after adoption of Bt. cotton. 58.94% of respondents said that their yields have gone up after adopting Bt. cotton (Thutupalli, 2015).

Further, India's share in the world cotton production increased to 25% in 2013 (Srivastava and Kolady, 2016). However, Bt. cotton is not free from criticism. The opponents alleged that secondary pests are emerging because Bt. cotton is no longer resistant. It was stated that animals died because of Bt. Toxin. It was even alleged that farmers committed suicides because of Bt. cotton crop failure. However, empirical evidence either finds no evidence on farmer suicides and Bt. cotton adoption or remains inconclusive (Gutierrez *et al.*, 2015).

8.4 Bt brinjal, an abortive move

Bt brinjal was the second GM crop officially approved for commercialisation in India but withdrawn immediately following nationwide uproar against its cultivation. Brinjal (eggplant) is one of the important vegetable crops in India accounting for 8% of the total production of major vegetables grown in the country. Fruit and Shoot Borer is the most destructive and unmanageable pest of brinjal causing yield loss of 60-70%, even after repeated insecticide sprays. Bt brinjal was developed by Monsanto, Mahyco combine for resistance to fruit and shoot borer. The technology involves use of *cry* gene obtained from *Bacillus Thureingensis*. The technology is said to have great advantage in terms of avoiding insecticide and crop losses. However, its release was opposed by a cross section of people in the society on the ground that it is not safe for human consumption and that it would affect the biodiversity of the crop for which India is centre of origin. The official release and withdrawal of the approval of Bt brinjal in 2009-2010 ushered in a new debate on GM crops in the country. The developments that followed i.e. setting up Parliamentary Standing Committee on GM crops, Technical Expert Committee (TEC) by Supreme Court resulted in moratorium on GM crops in India and demoralisation of scientific community and lack of interest among technology developers both in public and private sectors.

8.5 Other crops in the pipe line

However, the scene seems to be reversing with right wing political party coming to power since May 2014. In July 2014, 21 new varieties of genetically modified crops were approved for field trials (Table 8.1). But the Minister concerned quickly added that their government is cautious to give approval for commercialisation. He said: '*there are doubts in public minds over the impact of GM crops. So we are not in a hurry to accord sanction for their commercialisation. We have given a green light for field trials and scientific study of GM crops. Such trials and research will take 8 to 10 years. After the outcome of trials and research some decision will be taken'* (Javadekar, 2015). Thus, it is unclear as to which way the pendulum will swing after 8 to 10 years. As of now it looks uncertain though permissions are given for field trials.

Table 8.1. Transgenic crops under development and field trials (adapted from Department of Biotechnology, Government of India).

Crop	Organisation	Gene
Brinjal	IARI, New Delhi	*Cry1Ab*
	MAHYCO, Mumbai	*Cry1Ac*
Cauliflower	MAHYCO, Mumbai	*Cry1Ac*
	Sungrow Seeds Limited, New Delhi	*Cry1Ac*
Cabbage	Sungrow Seeds Limited, New Delhi	*Cry1Ac*
Chickpea	ICRISAT, Hyderabad	*Cry1Ac, Cry1Ab*
Groundnut	ICRISAT, Hyderabad	*IPCVcp, IPCV replicase*
Maize	Monsanto, Mumbai	*CP4 EPSPS*
Mustard	IARI, New Delhi	*CodA, osmotin*
	NRCWS, Jabalpur	*Bar, barnase, barstar*
	TERI, New Delhi	*Ssu-maize, Psy, Ssu-tpCrtl*
	UDSC, New Delhi	*Bar, barnase, barstar*
Okra	MAHYCO, Mumbai	*Cry1Ac*
Pigeonpea	ICRISAT, Hyderabad	*Cry1Ab+SBTI*
	MAHYCO, Mumbai	*Cry1Ac*
Potato	CPRI, Simla	*Cry1Ab*
	NCPGR, New Delhi	*Ama-1*
Rice	Directorate of Rice Research, Hyderabad	*Bacterial blight resistant, Xa-21, Cry1Ab, gna gene, sheath blight resistant*
	Osmania University, Hyderabad	*Gna*
	IARI, New Delhi	*Cry1Ac*
	MAHYCO, Mumbai	*Bt, chitinase, Cry1Ac and Cry1B-Cry1Aa*
	MKU, Madurai	*Chitinase, B-1, 3-glucanase, osmotin*
	MSSRF, Chennai	*genes from mangrove species*
	TNAU, Coimbatore	*Chitinase*
Sorghum	MAHYCO, Mumbai	*Cry1Ac*
Tomato	MAHYCO, Mumbai	*Cry1Ac*
	NCPGR, New Delhi	*OCDC*

8.6 Genetic modification controversy and public engagement in India

Like in many countries the debate on GM crops in India is highly polarised around pros and cons of the technology. People who support and oppose GM crops are divided on political party lines and ideological grounds. The GM debate in the country can be seen in three different phases. The first phase was when Bt cotton was released in 2002. Although there was severe opposition, Bt cotton was officially released in 2002 in a limited area. Very soon it spread throughout the country and at present more than 90% of cotton grown in the country is GM crop.

The second phase of GM controversy was when Bt brinjal (eggplant) was announced for commercial release in 2010. The opposition was so strong that the government slapped moratorium on Bt brinjal. This time around the government was directly involved in public engagement through multi-stakeholders consultative meetings. The then Minister of Environment and Forests under whose Ministry the apex approval committee (Genetic Engineering Approval Committee) operates, conducted these consultations. This is first of its kind in India to take public into confidence before a decision of national importance was taken. Public consultative meetings involving multi-stakeholders were held in seven major cities in the country. As many as 5,920 persons participated in the debate on pros and cons of Bt brinjal. Of them, about 3,000 were (50.6%) farmers and 1,065 (18%) were NGOs and 569 (10%) were scientists (Kalle, 2015). Majority of the participants in the debate voted against Bt brinjal and their main concerns were that it was not safe for human health and the regulatory process followed was inadequate and inappropriate. Following this, the Ministry decided to clamp a moratorium on not only on Bt brinjal but also other GMO crops which were under field trials.

The Union Government also gave powers to state governments to veto transgenic crop field trials. The decision had far reaching implications affecting the morale of the scientists and enthusiasm of both public and private research institutions.

8.7 Genetic modification debate and involvement of constitutional bodies

The debate on GM crops became so intense that the constitutional bodies such as the parliament and the supreme court of India were forced to get into the discussion.

8.7.1 Parliamentary Standing Committee

Following the division of political parties on GM crops in parliament, it was decided to examine in greater detail the pros and cons of the technology. Accordingly, a Standing Committee headed by Mr. Basudeb Acharia (Communist Party of India) was constituted. This Committee representing different political parties in the country made a thorough enquiry and presented its report to the Parliament on 9th August, 2012. The report assesses the benefits and drawbacks of the introduction of genetic modification in food crops. A major finding of the report was that the regulatory framework for GM crops has several shortcomings. It also noted that the current framework does not provide for mandatory consultations with state governments or seek their permission to conduct open field trials on GM crops, such as Bt cotton and brinjal. In light of these findings, the report recommended that all research and development activities on transgenic crops be carried out only in laboratories and that ongoing field trials in all states be discontinued (Balani, 2015).

8.7.2 Intervention of apex court

The Supreme Court of India also got involved in the debate through a Public Interest Litigation Case. In response to this, the Supreme Court constituted a Technical Committee to advise the court on GM crops. The report of the Technical Committee was submitted in July 2013 and it recommended an indefinite moratorium on field trials of GM crops till the government fixes regulatory and safety aspects and a ban on introductions of GM varieties in regions of their origin Chauhan (2013).

It may be interesting to note that the final report had a dissent note from Dr. R.S. Paroda, a well-known agricultural scientist representing the Ministry of Agriculture, Government of India on the TEC. The report is now with Supreme Court of India and a final decision is yet to be taken.

The third phase of the debate can be seen in the post 2014 after a right wing party won the general elections with an impressive mandate. The government in power quietly changed the course on GM field testing despite the fact that the decision of the apex court is still pending. Eight Indian states largely aligned with ruling Bharatiya Janata Party (BJP) have also approved field trials of GM crops.

Such a decision is reviving the enthusiasm of Indian biotech researchers. According to Dominic Glover at the University of Sussex, Brighton, UK 'India's attitude towards transgenic crops has symbolic importance beyond its borders because it epitomizes tensions that surround the use of the GM technology in developing nations' (Kumar, 2015).

The supporters of the GM technology in the present phase argue that India needs this technology more than the developed countries because of the compulsions the country faces. India needs to improve agricultural productivity to meet the growing needs of increasing population. India is also one of the seriously affected countries of climate change. GM technology has potential to offer solutions to climate resilient agriculture. Hence, the need for GM technologies, according to the supporters of the technology.

8.8 Genetic modification technology and regulatory mechanism

Being a signatory to International Conventions such as Convention on Biological diversity, India follows world standards in regulating the development and application of GM technologies. The GMOs and products thereof are regulated by regulatory framework notified under Environment

Table 8.2. Competent authorities and their functions in regulating GMOs.

	Name of the authority	Nature of function
1	The Recombinant DNA Advisory Committee	Advisory
2	Institutional Biosafety Committee	Approval
3	Review Committee on Genetic Manipulation	Approval
4	Genetic Engineering Approval Committee	Approval
5	State Biotechnology coordination Committee	Monitoring
6	District Level Committee	Monitoring

(Protection) Act 1986, referred as Rules 1989. According to these rules, there are six competent authorities under three categories of institutions as presented in Table 8.2.

- Genetic Engineering Appraisal Committee (GEAC) – functions under the Ministry of Environment and Forests. It is the apex body to accord environmental approval of activities involving large scale use of hazardous micro-organisms and recombinants in research and industrial production. It is also mandated with approving the release of genetically engineered organisms and products into the environment, including experimental field trials.
- Review Committee on Genetic Manipulation (RCGM) – functions under the Department of Biotechnology (DBT), Ministry of Science and Technology. RCGM is mandated with monitoring and regulating safety related aspects of ongoing research projects and activities, including small scale field trials.
- Recombinant DNA Advisory Committee (RDAC) – operates under the DBT, functions are mostly advisory in nature. It reviews developments in biotechnology, nationally and internationally.
- State Biosafety Coordination Committees (SBCC) – tasked with monitoring at the state level. It has the power to investigate and take punitive action in case of violations of statutory provisions.
- District level Committees (DLC) – responsible for monitoring at the district level.
- Institutional Biosafety Committee (IBSC) – The Committee is established under the institution engaged in GMO research. It oversees this research and acts as an interface between the institution and RCGM.
- Of these committees, the GEAC and the RCGM are the most crucial in the regulatory chain.
- 8.9 Procedure for approval of genetically modified organisms
  1. Initially, the company/agency involved in the development of the GM crop undertakes several biosafety assessments including, environmental safety, food and feed safety assessments in containment.
  2. This is followed by Bio-safety Research Trials in two stages Biosafety Research Level – (BRL) trial I and BRL-II which require prior approval of RCGM and GEAC respectively.
  3. Approval for environmental release is accorded by the GEAC after it considers the findings of the bio-safety and agronomic studies as well as recommendations of the RCGM and other committees.

Finally, commercial release is permitted by the GEAC for only those transgenic crops that are found to be safe for humans and the environment.

Procedures involved in the approval of GM crops in India are summarised in Figure 8.1.

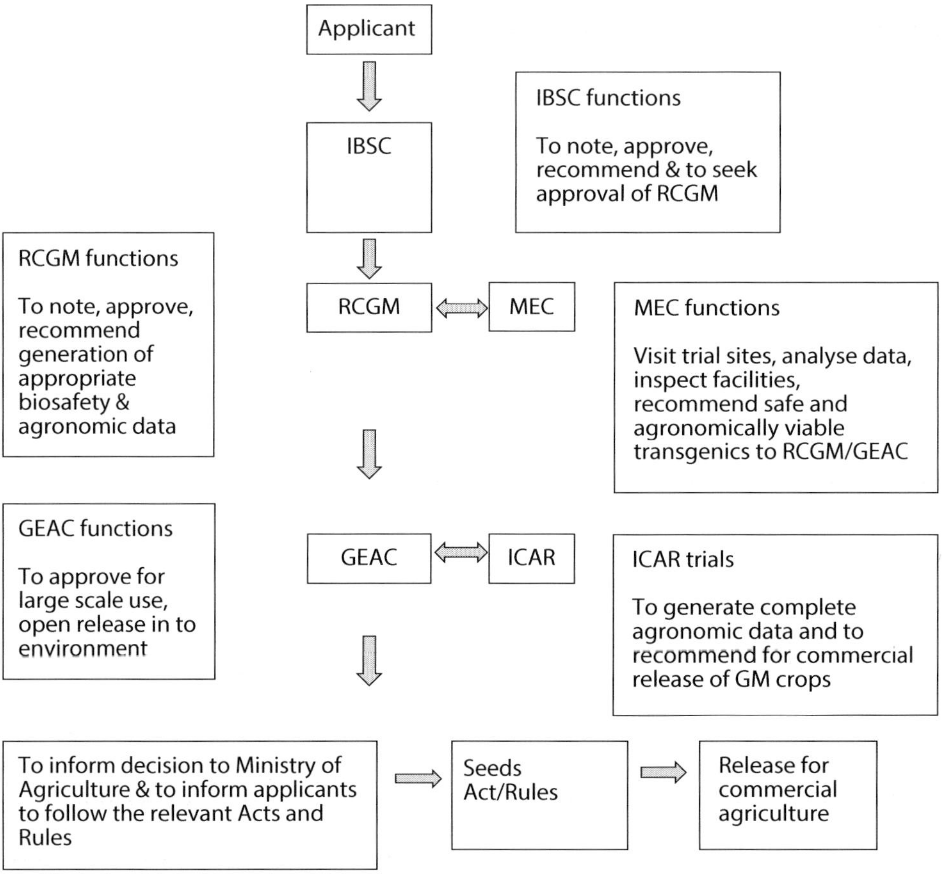

Figure 8.1. Approval process of GMOs in India. IBSC = Institutional Biosafety Committee; RCGM = Review Committee on Genetic Manipulation; MEC = Monitoring and Evaluation Committee; GEAC = Genetic Engineering Appraisal Committee; ICAR = Indian Council of Agriculture Research.

8.10 Biotechnology Regulatory Authority of India

During the first decade of commercial release of Bt. cotton, a number of lessons were learnt and certain limitations of the existing system were noticed. The Bt brinjal fiasco was also because of lack of confidence in regulatory procedures adopted. There is, therefore, an attempt to bring in a new legislation to create a more robust system of regulation in the form of Biotechnology Regulatory Authority of India (BRAI). The new legislation is pending before the Indian Parliament for the last three years. It is hoped that as and when approved it would offer a better and effective regulatory system. The proposed authority will be structured as noted in Figure 8.2.

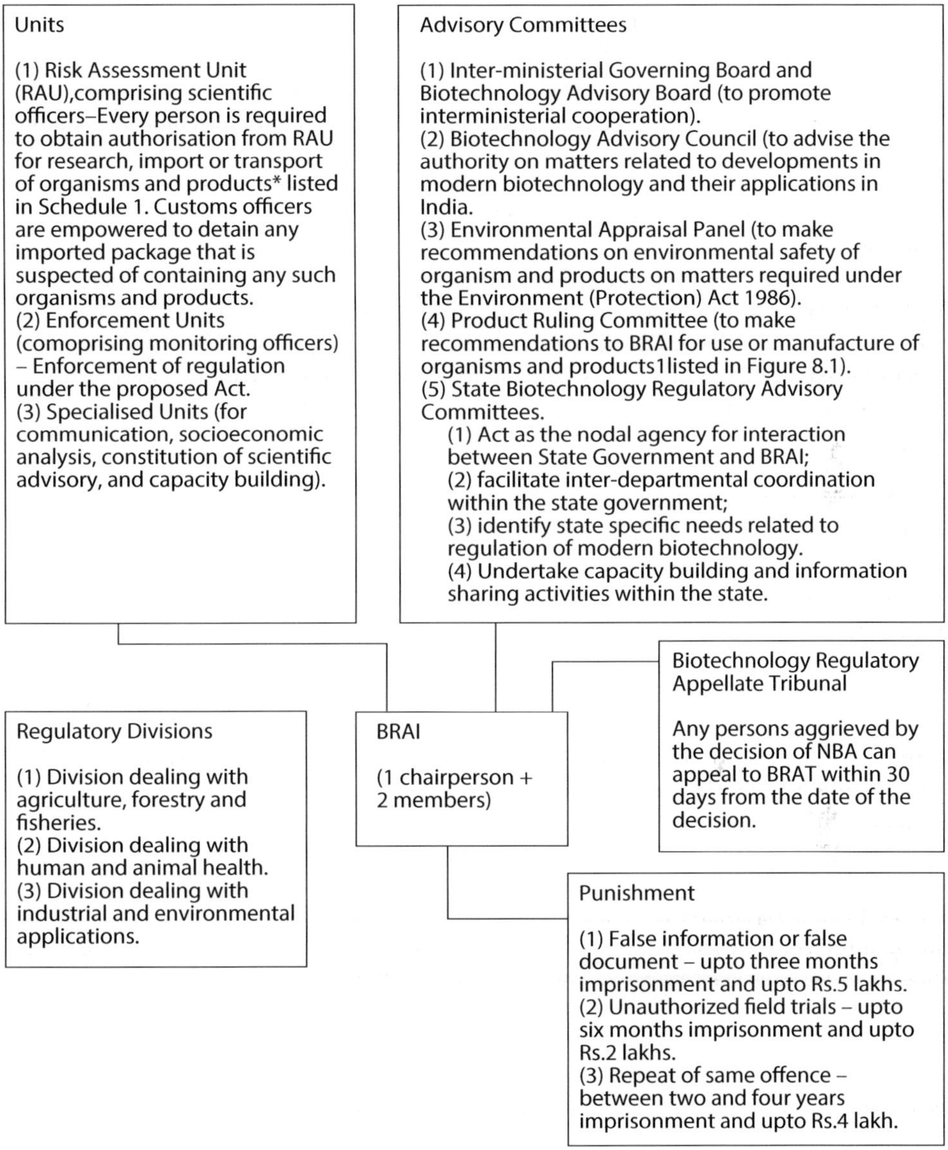

Figure 8.2. Structure of the Biotechnology Regulatory Authority in India. BRAI = Biotechnology Regulatory Authority of India; NBA = National Biosafety Authority.

[1] These include genetically engineered plant, animal, micro-organism, virus or other organisms that may have application in agriculture, fisheries, forestry or food production and that may be used as food; any animal clones that may have application in agriculture, fisheries and food production, recombinant proteins and combinations, vaccines containing living modified organisms, recombinant gene therapy products, etc. and any genetically engineered plant, animal, micro-organism, virus or other organisms that may be released into the environment or have application in industrial production.

8.11 Conclusions

India entered into cultivation of GM crops in 2002 after six years of developed countries produced GM crops. Although the first GM crop Bt cotton was introduced in the midst of strong opposition, very soon the controversy died down as the technology was widely accepted by farmers. India emerged as the leading producer of Bt cotton in the world and occupying 4th position among countries cultivating GM crops. However, it has not been smooth to release other crops for commercial cultivation. Bt brinjal which was officially announced for release was quickly withdrawn following public outcry. The debate took altogether a different turn with this edible crop compared with previous fibre crop where in even constitutional agencies like Parliament and Supreme Court directly got involved in the discourse. The net result was a moratorium on technology for about five years which impacted the morale of scientific community and research organisations. However, the scene got reversed since 2014 when a right wing political party captured power at national level. The apex approval agency which was almost defunct for about three years became active again and fresh approvals were given for 21 crops for field trials. Although a legal decision is pending at the apex court of India, the move by government giving fresh approvals sends a strong positive signal that India is again moving towards GM crops. However, it will be quite some time before some of the crops are released for commercial cultivation. In the meantime it is hoped that a more robust new regulatory system will be installed for speedy and effective decision making and better monitoring of GM technology in the country.

References

Balani, S. (2015). Standing committee report summary. Cultivation of genetically modified food crops – prospects and effects. PRS Legislature Research, New Delhi, India. Available at: http://tinyurl.com/jz9rhvu.

Chauhan, C. (2013). SC committee says no to GM crops for time being. Hindustan Times, July. Available at: http://tinyurl.com/zatueqo.

Government of India (2015). National biotechnology development strategy 2015-2020. Ministry of Science and Technology, Department of Biotechnology, New Delhi, India.

Gutierrez, A.P., Ponti, L., Herren, H.R., Baumgartner, J. and Kenmore, P.E. (2015). Deconstructing Indian cotton: weather, yields, and suicides. Environmental Sciences Europe 27(12).

International Service for the Acquisition of Agri-biotech Applications (ISAAA) (2014). Global status of commercialized Biotech/GM crops: 2014. ISAAA Brief 49-2014, Executive Summary. ISAAA, New York, NY, USA.

Javadekar, P. (2015). Message of the Union Environment Minister, 9th June, 2015 (press release).

Kalle, J. (2015). Post normal science perspective of genetically modified (GM) food controversy. The case of BT brinjal in India. Unpublished Ph.D. thesis, University of Hyderabad, Hyderabad, India.

Kumar, S. (2015). India eases stance on GM crop trials. Nature 521: 138-139.

Narayanamoorthy, A. and Kalamkar, S.S. (2006). Is Bt. cotton cultivation economically viable for Indian farmers? Economic and Political Weekly 41(26): 2616-2724.

Srivastava, S. and Kolady, D. (2016). Agricultural biotechnology and crop productivity: macro-level evidences on contribution of Bt. cotton in India. Current Science 110(3): 311-319.

Thutupalli, A. (2015). Technology paradigm shifts in agriculture: drivers of sustainability and catch up. Ph.D. thesis, Maastricht University, Maastricht, the Netherlands.

9. GMO risk assessment in the EU: interplay between science, policy and politics

J.M. Casacuberta[1*], *F. Nogué*[2], *P. du Jardin*[3]
[1]*Center for Research in Agricultural Genomics, CRAG (CSIC-IRTA-UAB-UB), Campus UAB, Cerdanyola del Vallès, 08193 Barcelona, Spain;* [2]*INRA AgroParisTech, IJPB, UMR 1318, INRA centre de Versailles, route de Saint Cyr, 78026 Versailles CEDEX, France;* [3]*Université de Liège, Gembloux Agro-Bio Tech, 2 Passage des Déportés, B-5030 Gembloux, Belgium*[3]*; josep.casacuberta@cragenomica.es*

Abstract

This chapter analyses the evolution of GMO risk assessment in the EU from the first discussions within the scientific community until now, situating as landmarks in the analysis the reforms that have taken place within the legal framework. After an introduction there is a presentation of the key elements of GMO risk assessment in the 1990s directives, which typically situated Member States as crucial components of the system. This is followed by analysis of the adoption of the second regulatory framework for GMOs in the EU, the keys of which were Directive 2001/18/EC, Regulation 1829/2003 and the creation of the European Food Safety Authority. The third section details the significant differences that in the application of this second legal framework begin to appear in the 2000s between imported GM products and GMO cultivation in the EU. This analysis is followed by a discussion, finally and prior to the conclusions, of how with regard to risk assessment the balance between the roles of the different actors, as well as between different values at stake (such as scientific soundness, independency, transparency, consistency) has evolved.

Keywords: EFSA guidance for GMOs, risk assessment, scientific soundness, independency, transparency, consistency, genome editing

9.1 Introduction

The great potential for plant breeding of genetic transformation was clear from the first reports on gene transfer to plant cells in the laboratory (Barton *et al.,* 1983; Herrera-Estrella *et al.,* 1983). Transgenesis was opening new ways to modify genomes and was greatly increasing the palette of possible traits to be introduced in plants, and this rapidly translated into projects to commercialise transgenic plants with new traits. Whereas plant breeders and seed companies were enthusiastic about the potential of these techniques, the possible risks associated with the release of GMOs were rapidly put forward and different environmental groups tried to block the development of these techniques (Fox and Norman, 1983; Norman, 1983; Sun, 1984). This debate was rapidly followed by discussions within the scientific community (Brill, 1985; Kolata, 1985) and prompted a discussion on the legal framework of GMO commercialisation and the establishment, well before the first product obtained from transgenic plants was ready for commercialization, of specific legislation in both the US and the EU. This first legislation on the deliberate release into the environment of GMOs in the EU, the Council Directive 90/220/EEC (Council of the European Communities, 1990), has been complemented and modified over the years. The analysis of the potential risks associated to GMOs is at the heart of all those legal texts, and all these texts contain principles that have remained

DOI 10.3920/978-90-8686-845-2_9., © Wageningen Academic Publishers 2017 unchanged, such as the need to have a case-by-case evaluation of those potential risks. However, the role of the different European and EU Member State (MS) authorities and scientific experts in the authorization procedure, the organization of the process itself, as well as the actual requirements for the risk assessment and authorization of GMOs have changed substantially over the years. Moreover, in addition to the Directives and Regulations issues by the EU Council and Commission, other texts, such as the European Food Safety Authority (EFSA) Guidances have completed the framework in which the risk assessment is performed in the EU. The aim of this article is to analyse the evolution of the GMO risk assessment in the EU, and more precisely to discuss how the balance between the roles of the different actors, as well as between different values at stake (such as scientific soundness, independency, transparency, consistency) has evolved.

9.2 The early days: the Member State at the core of the system

The first GMO legal framework for the deliberate release into the environment of GMOs in the EU was established in 1990 with Directive 90/220/EEC. This Directive already contained some of the basic principles that still inspire the GMO risk assessment in the EU. The Directive defines what is legally considered as a GMO and lists the techniques which lead to a GMO and those which do not. This therefore places an emphasis on the methods used to obtain a product, and represents one of the main differences in GMO regulation in the EU compared with other countries and regions, such as the USA, where the focus is more on the nature of the final product (Lynch and Vogel, 2001). The core of the Directive is the analysis of the risks associated with the deliberate release of GMOs and the placement on the market of GMOs and derived products. It urges each MS to designate a Competent Authority (CA) to carry out the requirements established in the Directive. The risk assessment of a new GMO is then performed by the CA of the MS where the deliberate release, or the placing on the market, would take place. In cases where no risks are identified, the CA of the MS can then allow the deliberate release, or the placing on the market, of the GMO or its derived products to take place in the EU. The MS carrying out the risk assessment then informs the European Commission and the other MS of its decision. However, one or more MS can object to this decision, and where a consensus cannot be reached, the final decision becomes that of the Commission.

Therefore, Directive 90/220/EEC left both the risk assessment and the authorisation decision in the hands of the CA designated by the corresponding MS. In this system the risk assessment is provided by experts directly linked to governmental bodies of each MS. It was only in 1997 that the European Commission decided to create a centralised scientific assessment system through a number of Scientific Committees in the field of consumer health and food safety. One of these Committees, the Scientific Committee on Plants, was in charge of scientific and technical questions on human or animal health or the environment relating to GM plants intended for human or animal consumption or for production or processing of non-food products (Commission of the European Communities, 1997). The intention of the Commission Decision was to increase food safety and public trust by ensuring a mechanism to obtain scientific advice 'based on the principles of excellence, independence and transparency'. For this reason, clear rules on how to select the experts members of the Panels, and how to declare any possible interest that may lead to a conflict of interest were established. In parallel the agendas of the Committee were made public together with the written scientific opinion and any other information important for transparency purposes.

This was the first step towards a modification of the GMO risk assessment and the GMO legal framework in the EU which at that time was perceived to be inadequate and that was heavily questioned by different environmental and political groups (Lee, 2008). Several food crises hit the EU in the 1990's, including the Bovine Spongiform Encephalopathy (BSE) crisis, which contributed to mistrust of the existing food safety system. On the other hand, media coverage of the first cloned animal, Dolly the sheep in 1996, stimulated discussions on the safety and the ethical implications of the use of biotechnology on domestic plants and animals, and also contributed to strengthen the idea that a tighter GMO regulation was needed. Finally, in 1998, Dr Arpad Pusztai announced on a television program that rats fed on GM potatoes exhibited problems with body weight and immune system responses. When this experiment was finally published (Ewen and Pusztai, 1999), it received strong criticism from the scientific community that considered the design flawed and the conclusions not supported by the data (Lemaux, 2008), and Dr. Pusztai was even suspended from his post as research scientist. This was one of the different cases that had impacted significantly on the public perception of GMOs and on how their potential associated risks were assessed in the EU.

As a consequence, the European Council requested the European Commission to adopt more stringent and transparent rules for the placing on the market of GMOs and to put in place mandatory rules on labelling and traceability.

9.3 The intermediate phase: Directive 2001/18, Regulation 1829/2003, and the creation of EFSA

Between 1998 and 2004 no applications for authorisation of GMOs were approved, in what was called a *de facto* moratorium. During this time the entire GMO regulatory framework was revised, with a new Directive in 2001 for the deliberate release of GMOs (Directive 2001/18/EC), the Regulation on GM food and feed (Regulation 1829/2003) and subsequently in 2003 the Regulation concerning the labelling and traceability of GMOs (Regulation 1830/2003). In 2003 EFSA was also created to provide scientific advice to the EU institutions in areas related to food safety thereby replacing the different Scientific Panels, including the Scientific Committee on Plants, created some years earlier. This was a new attempt to increase food safety and public trust in the system through the creation of an independent agency within which external scientific experts organised into different scientific panels became responsible for issuing scientific opinions on food safety-related issues. One of these Panels, the Panel on Genetically modified organisms ('GMO Panel') provides independent scientific advice on food and feed safety, and environmental risk issues related to GMOs. In addition to the external scientific experts members of the different Panels, EFSA employs its own scientists to support the work of the Panels.

The changes in the legislation coupled with the establishment of EFSA clearly resulted in a more centralised and independent risk assessment process. Under the Directive 90/220/EC the risk assessment and the authorisation decision was the responsibility of a MS, although the rest of the MS and ultimately Commission also participated in the process. Under Regulation 1829/2003 the core of the risk assessment is transferred to EFSA, and the decision on adoption or rejection of applications for GMO commercialisation is taken collectively by all MS or ultimately by the Commission. As a consequence, the risk assessment and the approval of GMOs are more clearly separated which help to ensure both the scientific excellence and the independence of the scientific advice.

The changes introduced in the legislation not only modified the procedures but they also slightly changed the principles of the risk assessment and to some extent the requirements. Both the Directive 2001/18 and the Regulation 1829/2003 placed a stronger emphasis on the precautionary principle as the basis of the legal framework, and stressed the need to consult the public before any authorisation is granted. In addition, both the Directive and the Regulation limit the authorisation of the deliberate release or the placing on the market of a GMO to a period of 10 years after which a renewal of authorisation is required. Moreover, Directive 2001/18 establishes a procedure, called the safeguard clause, to provisionally prohibit the commercialisation or use of a GMO in its territory, in case new information becomes available which indicates a possible risk. Finally, the legislation put in place during this period establishes clear labelling and traceability rules for approved GMO products and describes a strong monitoring strategy that reinforces the control over the approved GM plants and derived products.

With respect to the risk assessment itself, whereas the Directive 2001/18/EEC confirms that the case-by-case analysis (i.e. the variation of the required information depending on the type of the GMOs concerned, their intended use and the potential receiving environment) is the principle for the risk assessment it also states that harmonised procedures are needed. This justifies more detailed annexes to the Directive where the principles and the information required in the application for a GMO authorisation are laid down. However, neither Directive 2001/18/EEC nor Regulation 1829/2003 contained precise data requirements and methods to obtain these data. For this reason, and in order streamline the procedure and ensure consistency in the risk assessment of the information received from different applicants, in 2003, and before EFSA was created, the Joint Working Group on Novel Foods and GMOs prepared for the Scientific Steering Committee a Guidance for applicant describing the information required (Joint Working Group on novel foods and GMOs, 2003). This Guidance formed the basis of the EFSA Guidance document for the risk assessment of GM plants and derived food and feed products that was published in 2006 (EFSA, 2006b) and which contained more precise indications on the data required from applicants and also on methodologies and approaches used to collect these data. However, in some aspects these indications still remained quite general. The document was indeed used only as guidance by the GMO Panel, it was not used to prescribe a list of mandatory requirements, and the principle of the case-by-case approach justified some latitude in applying the elements of this guidance. This first EFSA GMO Guidance published in 2006 was updated several times in subsequent years with additional, more specific information included on requirements for risk assessment. This has been complemented by the publishing of subject specific guidance documents. In particular, the GMO Panel published a Guidance on applications to renewal the authorisation of GMOs in 2006, a Guidance on applications on stacked events in 2007 and a Guidance on GMOs for non-food non-feed uses in 2009 among others (EFSA, 2006a, 2007, 2009d).

9.4 From guidance to legal rules: the Commission implementing Regulation 503/2013

More than 30 GMOs (maize, soybean, oilseed rape, cotton and sugar beet) and derived products were approved for import under the new regulatory and risk assessment framework. However, the end of the *de facto* moratorium for imported products was not extended to applications for cultivation in the EU, with the exception of the placing on the market of a modified starch potato in 2010 which was afterwards withdrawn by the applicant, and the pest resistant GM maize MON 810 is still today the only GMO cultivated in the EU. In fact, several EU MS invoked the Safeguard clause of the Directive 2001/18 to ban the cultivation of MON 810 in their territory. France was the first to invoke it early in 2008 based on a report of a national committee. This was itself controversial as twelve French scientists and two economists opposed to the conclusions drawn (GMO Compass, 2008). EFSA reviewed the studies on which the application to invoke the Safeguard Clause was made and concluded that they did not support scientifically the decision to ban cultivation of MON810 (EFSA, 2008a). Subsequently, both the European Court of Justice and the French Conseil d'État concluded that there was no proof of the existence of a particular high level of risk justifying the ban, which was therefore declared as illegal (http://tinyurl.com/mnh8vll). Hungary and Greece also invoked the Safeguard clause in 2008 to ban the cultivation of MON 810 and, as in the case of the French ban, these measures were also found to be unfounded scientifically by EFSA (EFSA, 2008b,c)

The Safeguard clause was also invoked by Austria to ban the import of approved GMOs, such as maize MON863, and oilseed rapes GT73, MS8, RF3 and MS8xRF3. Although these measures were also found to be unfounded scientifically by EFSA (EFSA, 2009a,b,c), the situation created by the recurrent use of the Safeguard clause to ban the commercialisation of approved GMO products, showed that at least some MS were clearly not satisfied with the existing risk assessment and the regulatory framework in place. There were multiple reasons for this. On the one hand, regulation 1829/2003 became the entry door for most GMO applications, thus the role of MS in the risk assessment and the authorisation process, and hence in controlling the entire system, was reduced (Box 9.1). Whilst MS are consulted and can send questions on the risk assessment dossiers to EFSA, it is EFSA who requires the applicant to provide specific data and who eventually produces the corresponding scientific opinion. In addition, although the decision on the possible authorisation of a GMO is first at the Standing Committee on the Food Chain and Animal Health, where the MS sit, in almost all cases the lack of a qualified majority vote for or against approval for commercialisation leaves the final decision to the EU Commission.

Box 9.1. Authorisation procedures of GMOs under Regulation (EC) No 1829/2003 and under Directive 2001/18/EC, highlighting the role of MS during the risk assessment (adapted from EFSA; http://tinyurl.com/k8j59qo).

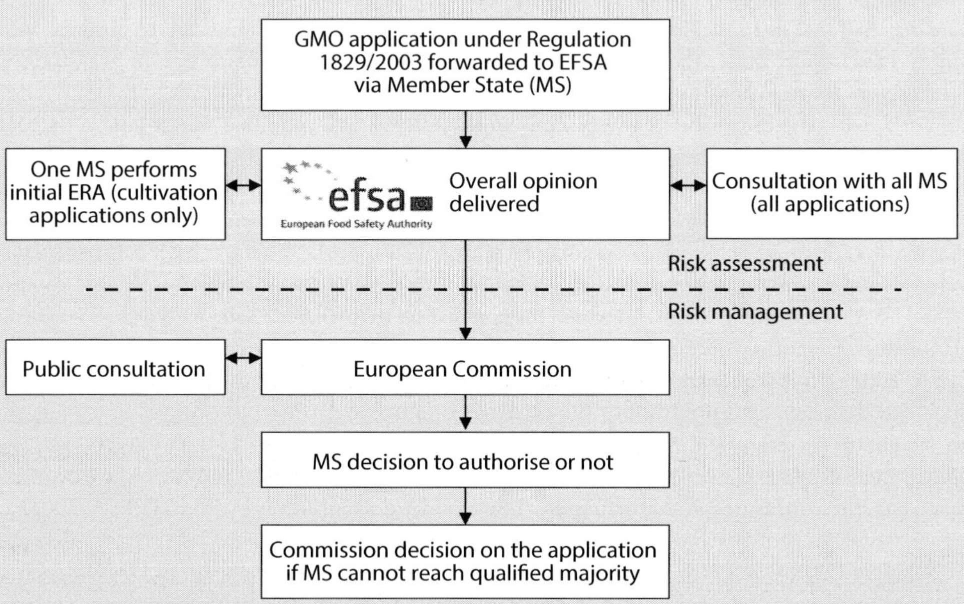

Figure 9.1. Authorisation procedure under Regulation (EC) No 1829/2003 (centralised procedure).

Directive 2001/18/EC, which is valid for all scopes of applications of GMOs, designates one MS as rapporteur of the risk assessment. All other MS evaluate the dossier as well and have the capacity to raise objections to the conclusions of the rapporteur MS. The EC examines whether the objections can be resolved or not. If not, EFSA is mandated to evaluate the dossier and to issue its own conclusions in a scientific opinion delivered to the EC and MS. These latter are responsible for taking the decision on the applicant's request by qualified majority voting. Directive 2001/18/EC (Figure 9.2) recalled the Directive 1990/220/EC, which did proceed in a similar way, but without EFSA a second risk assessment body in the procedure.

Regulation (EC) No 1829/2003 applies to food/feed applications of GMOs, whether the GMOs are intended to be cultivated in the EU or not and states a centralised authorisation procedure (Figure 9.1). For 'import dossiers' (i.e. no cultivation foreseen in the EU), EFSA performs the initial risk assessment on all aspects of the application. For cultivation dossier, a MS is designated as rapporteur to perform a preliminary environmental risk assessment, the conclusions of which are passed to the GMO experts panel of EFSA which is responsible for issuing the final risk assessment report in a scientific opinion published in the EFSA Journal. For both import and cultivation dossiers, all MS are consulted during the risk assessment and the EFSA GMO panel must respond in writing to any comments and objections within its remit, but a consensus is not searched for. Based on the EFSA opinion, the EC drafts a proposal of decision which is submitted for qualified majority vote to the MS.

The comparison of both procedures (Figures 9.1 and 9.2) highlights that the risk assessment under the Regulation (EC) No 1829/2003 gives EFSA the prominent role in the risk assessment, the MS being only consulted. In the previous directive 90/220/EC, and to a lesser extent in the current directive 2001/18/EC, the MS had/have a more prominent role (although EFSA is given the final say in case of unresolved objections during the risk assessment under directive 2001/18/EC).

Since its adoption, the Regulation (EC) No 1829/2003 was applied in preference to virtually all food/feed dossiers, the directive being applied only in the few cases of non-food/feed dossiers submitted by the applicants. As a conclusion, since the entry into force of the Regulation (EC) No 1829/2003, EFSA should be regarded as the leading authority in the risk assessment of GMOs in the EU, with a limited influence of the MS during the risk assessment phase of the authorisation procedure.

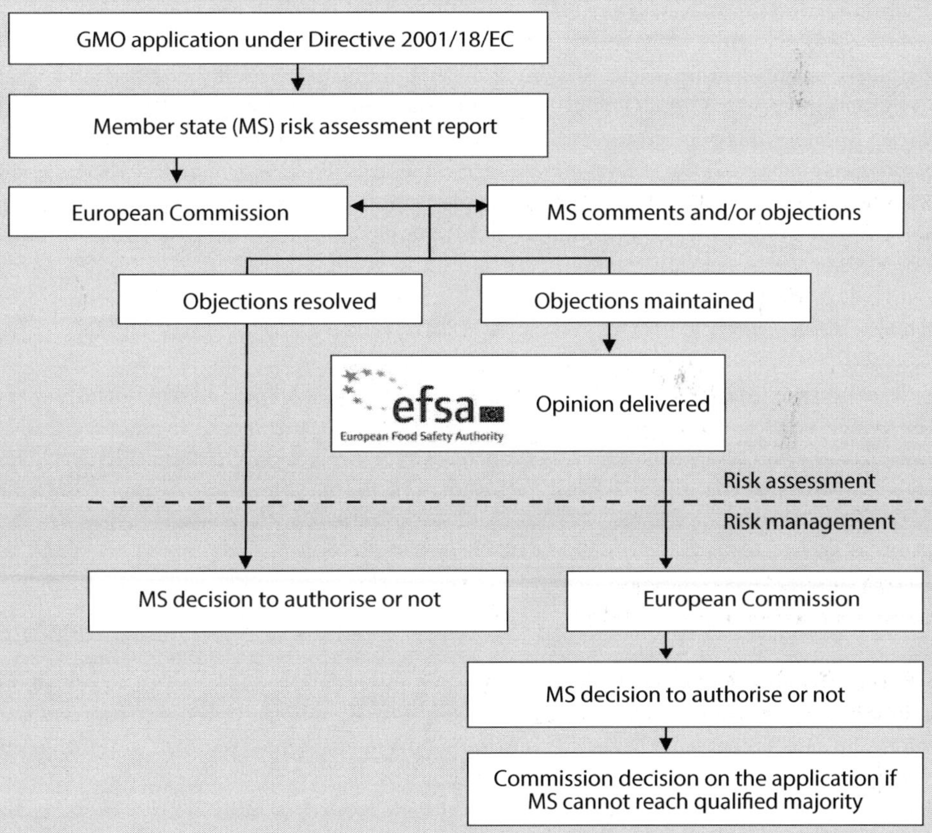

Figure 9.2. Authorisation procedure under Directive 2001/18/EC.

In addition, a number of scientific reports pointing to possible risks of some GM products were published (see Box 9.2 for a detailed example). The conclusions drawn were discussed and challenged in other scientific articles, by EFSA itself and by other scientific bodies (Box 9.2). The conclusion was that the presumptions of risk were not supported by the data presented. Noteworthy is the general consensus that the conclusions proposed in these articles were unsubstantiated, and despite that further publicly funded scientific projects failed to find the risks indicated in these articles, they were widely reported in the media fuelling the opposition of some MS to the risk assessment and regulatory framework in place.

Box 9.2. The Séralini 2012 paper.

Gilles-Eric Séralini is professor at the University of Caen. He is president of the scientific advisory board of the Committee of Research and Independent Information on Genetic Engineering (CRIIGEN), an organisation that has shown a strong opposition to genetically modified food. Séralini has published multiple studies claiming health risks associated with GMOs and the glyphosate-based herbicide Roundup. He published in 2012 an article titled: *Long term toxicity of a Roundup herbicide and a Roundup-tolerant genetically modified maize* in the revue Food and Chemical Toxicology. The article presented a two-year feeding study in rats, fed with Roundup-tolerant genetically modified maize, cultivated with or without Roundup, and Roundup alone reported an increase in tumours among rats fed with genetically modified corn and the herbicide RoundUp.

Reaction of the scientists

Rapidly many scientists questioned the methodology used by the authors and the conclusions of the study[1] even if few scientists, like Paul Deheuvels, for example, a member of the French Académie des Sciences, supported the conclusions of the article. In France, a report[2] from six French scientific academies concluded that the article was of low quality and questioned professional and ethical aspects of Séralini's work, including the fact that the author organised in advance intensive media coverage of his conclusions.

Reaction of the risk assessors

From all the national and supra-national food safety and regulatory agencies that evaluated the Séralini paper, the Belgian Biosafety Advisory Council (BBAC) was the only one to express a mixed opinion[3]. Indeed, 3 experts out of the 11 constituting the panel pointed out that 'the long duration of this study is a positive aspect since most of the toxicity studies on GMOs are performed on shorter periods' and in its conclusion the BBAC proposed EFSA '*urgently to study in depth the relevance of the actual guidelines and procedures*'[3].

All the others food safety and regulatory agencies including, the German federal institute for risk assessment[4], three Canadian regulatory agencies[5], the French HCB[6] the French food safety agency ANSES[7], the Vlaams Instituut voor Biotechnologie[8], the Technical University of Denmark[9], the bi-national (Australia and New Zealand) Government agency FSANZ[10] and the Brazilian National

Technical Commission on Biosafety[11], and the EFSA[12], concluded that the study's design was flawed and its conclusions unsubstantiated.

Reaction of the risk managers

The first reactions of the national representatives to the Séralini study were cautious and based essentially on the conclusions of the risk assessment agencies. Three countries had a different perception on the Séralini case. The first one, in France, where the Minister of Agriculture, said of the Séralini study that '*it reinforces the position that was ours*'[13] making reference to the proposition of a moratorium for all genetically modified crops in France. In addition, the Minister declared that the GMO approval procedures in the European Union should be reconsidered. The Séralini case had also probably to do with the decisions of the Russian government that, in late September 2012, temporarily suspended importing GM corn and of the Kenyan government, who imposed a ban on genetically modified foods in November 2012. Officially, the Séralini paper had no influence on the regulating legislation of the others national authorities.

Impact of the article in different fields

The scientific impact of the article in the field of food safety has been very poor and, following the general view in the scientific community that the Séralini paper did not deserve publication in a peer reviewed journal, the editor of *Food and Chemical Toxicology* retracted the paper in November 2013 after the authors refused to withdraw it. In June 2014 the article was republished including raw data of the original study, without peer reviewing, in the journal *Environmental Sciences Europe*. This journal has no Impact Factor, a measure reflecting the yearly average number of citations to recent articles published in an academic journal.

In spite of the low scientific impact, the article had a clear impact on European and French scientific project calls as two projects (GMO90+, total cost € 3,695,000 and G-TwYST, total cost € 3,773,000) related to the Séralini study have been launched since 2012.

Impact on the general public of the 2012 Séralini paper is difficult to estimate. However, the photographs from the journal article of treated rats with large tumours that were widely circulated in the press[14], and continue to be shown in the front-page of the CRIIGEN and other websites, with a clear intention to influence public perception of GMOs.

[1] Butler, D. (2012). Rat study sparks GM furore. Nature 489: 484.

[2] http://tinyurl.com/mb2qnrw.

[3] http://tinyurl.com/m8ppd6c.

[4] http://tinyurl.com/m4pfkaf.

[5] http://tinyurl.com/cxz4pus.

[6] http://tinyurl.com/m9deegn.

[7] http://tinyurl.com/lts6fcx.

[8] http://tinyurl.com/cyceyeh.

[9] http://tinyurl.com/lha5pv2.

[10] http://tinyurl.com/l294re2.

[11] http://tinyurl.com/m9rbwse.

[12] http://tinyurl.com/l8xqz42.

[13] http://tinyurl.com/8ebnk2n.

[14] http://tinyurl.com/k3tg2ct.

All these reasons explain why shortly after the entry in force of the Regulation 1829/2003, the Commission started to work with MS to amend the regulatory framework in place. MS were given more control over the scientific risk assessment performed by EFSA by defining the data requirements in a more specific and detailed way. As opposed to the general framework defined in the Regulation 1829/2003, complemented by different EFSA Guidances, the Implementing Regulation 503/2013 established in a legal document an extensive list of mandatory data requirements and of methods to be used for generating the data. The implementing Regulation 503/2013 was the result of political negotiations between MS and, although it was based on the existing EFSA Guidance, it incorporated data requirements (e.g. mandatory 90-days rodent feeding studies) that EFSA deemed unjustified and that were indeed heavily criticised by the scientific community (Devos *et al.,* 2016; Kuiper *et al.,* 2013). In this way, MS reduced the latitude of the EFSA GMO Panel scientific experts to ask for data on a case-by-case basis, and gained more control over the risk assessment process.

In addition to the major changes introduced by the Implementing Regulation 503/2013, the way risk assessment is performed was also influenced by several developments within EFSA itself. On one hand, whereas at the early years of EFSA almost all the risk assessment work, including the drafting of documents, was performed by the experts of the GMO Panel, EFSA's own scientific officers (which increased in numbers associated with the GMO Panel) became more involved in these activities. This was also accompanied by an increasing involvement of external contractors in the primary analysis of specific parts of the dossiers. It is also the case that the work of the GMO Panel was more and more framed by cross-cutting Guidances developed by EFSA, dealing with issues such as expert knowledge elicitation or statistical reporting.

9.5 Tensions and equilibriums between actors and values in the GMO risk assessment in the EU

It is clear from the above that we can identify a range of actors and values which contribute to strained relationships, or 'tensions', throughout the last 20 years of the evaluation of GMOs in EU. We propose that these tensions have driven significant changes in the regulation and risk assessment of GMOs in the EU, modifying the equilibrium between these actors and values. We have identified different tensions, among which some are purely political whereas others involve scientific, ethical and societal issues.

The first clear tension is of a political nature and it is not specific to the GMO risk assessment. The very nature of the EU, composed of an heterogeneous ensemble of MS that give to the EU the power to legislate and control certain areas, implies that there will always be a tension between the EU centralised power, represented by the European Commission, and the MS. This is the case for the GMO risk assessment and approval. Whereas in the 1990's the risk analysis and approval of GMO applications was fully in the hands of the MS, the Regulation 1829/2003, and to a lesser extent the directive 2001/18, moved the core of the risk assessment to EFSA, an independent and centralised risk assessment body. The present situation, in which EFSA has to perform the risk assessment within the framework of a tight Implementing Regulation politically negotiated and agreed by MS, takes back to the MS, albeit collectively, some control over the scientific risk assessment.

However, even before the entry into force of the Implementing Regulation, the influence of MS in the risk assessment of GMOs performed by EFSA increased over time. Whereas in the first years of EFSA only the questions and issues raised by MS on GMO dossiers that were considered as relevant by the EFSA GMO Panel were directly addressed in the opinion, EFSA was subsequently obliged to address one by one all questions in an annex to the opinion. Similarly, the use of the Safeguard clause by some MS to block the commercialisation of authorised products can also be seen as an intent to regain control over the risk assessment and authorisation of GM Plants and derived products. An interesting case that illustrates this strong tension throughout the years is the long and complicated case of the authorisation for cultivation of Pioneer 1507 maize. Since the submission of the application in 2001 the Spanish competent authority (in 2003) and EFSA (in 2005, 2008, 2011 and 2012) delivered positive opinions. However, the approval is still pending and under discussion (November 2016) due to the pressure of the MS and the European Parliament, and this even after a decision of the European Court of Justice urging Commission to take action in 2013 (General Court, 2013).

There are other tensions that are more specific to the GMO risk assessment. One of the most important is the tension between consistency and the case-by-case approach. The former ensures similar treatment of different applicants and dossiers and allows the system to be predictable, which is highly demanded by both risk managers and applicants. It also results in quality assurance which aims at securing, from a legal point of view, the scientific outputs of EFSA, and at strengthening its communication. On the other hand, the case-by-case approach allows adjustments in the assessment taking on-board, for example the GMO trait in question its intended use and the potential receiving environment, which is required for a scientifically-sound assessment. Whereas the case-by-case approach was introduced in the legislation as one of the pillars of the risk assessment of GMOs, the changes introduced over the years to ensure consistency has certainly allowed MS more control over the scientific risk assessment, but has clearly compromised this approach. The most clear example being the Implementing Regulation 503/2013 which, after political negotiations between MS, has fixed as mandatory the requirements and methods to perform the risk assessment, making it even illegal to deviate from it, even in case where there is no scientific justification to ask for a particular set of data.

A somehow related strain could be defined as the tension between professionalization of the risk assessment procedure and the independence and excellence of the scientific expertise. Risk assessment is not solely a scientific evaluation. It is embedded in policy and communication issues, designed to promote consistency, transparency and independence. Promotion of these values has increased the bureaucratic burden, called for by quality assurance procedures, development of communication skills, etc. As a consequence, the scientific expertise tends to professionalise, with an increasing responsibility assigned to in-house scientists of the risk assessment body. Another trend is the out-sourcing of certain tasks and studies allowing EFSA to keep control of what is done and communicated, and how. In this context, joining expert committees becomes less attractive for top scientists, who are asked to dedicate more time and efforts to develop skills and capacities which are hardly in line with their core scientific activities. The interest of a system based on external scientist on the forefront of the research disappears when they become professionals of the risk assessment with limited time dedicated to their research.

There is a tension that is inherent to the risk assessment, due to the probabilistic nature of the risk. This is difficult issue of dealing with uncertainty during the risk assessment and yet giving a clear message on the risk to risk managers, who have difficulties in dealing with uncertainty. This tension points to the difficulty of implementing the precautionary principle in the GMO regulation. The precautionary principle applies in situations of 'lack of scientific certainty'. In spite of multiple interpretations by politicians, scientists and lawyers, we can say that it asks the risk assessor to identify and quantify the uncertainties, to identify the sources of uncertainties and the ways to resolve them. The risk assessment report is transmitted to the risk manager, i.e. the decision-maker, who will decide how to deal with the uncertainties and how to communicate to the outside world (stakeholders, citizens) on both risks and uncertainties. In practice, GMOs have become a very sensitive case in the public, the media, and the politicians, and this results in the incapacity of risk managers to acknowledge and communicate on uncertainties. In the EU, where the EC has the capacity to mandate EFSA to help resolve food risk-related issues, the risk manager has been putting more and more pressure on risk assessors to clarify any uncertainty-related issue, asking for more detailed opinions and for filling knowledge gaps that create uncertainty.

Scientific risk assessment is an interdisciplinary activity at the interface of different disciplines and requires a balance of approaches and cultures not always easy to facilitate. Science is a cumulative activity that progresses by confronting newly generated data with the current consensus. A single publication challenging the consensus generates interest but does not lead necessarily to its revision. It is only after the scrutiny of the scientific community and when other scientists are able to obtain the same results independently, that the previous consensus is modified. On the contrary, and especially in controversial and sensitive areas like the risk associated with the food and feed chain, the importance of an isolated piece of information contradicting the current consensus and supporting the views of particular stakeholders can easily be magnified in the media. Risk assessment should be based on sound science, but at the same time has to provide useful input to risk managers and politicians who are under the scrutiny and the pressure of society and the media. Because of this pressure, some isolated articles which have implied a risk associated with a GMO have strongly influenced the GMO regulation, even when the scientific community unambiguously considers the conclusions of these articles not to be supported by the data presented (see Box 9.2 for an illustrative example). On the contrary, the important body of work, funded in many instances by the EU (e.g. the GRACE project, http://www.grace-fp7.eu/), and the continuous surveillance of GM plants over more than 25 years of commercialisation, which has failed to detect particular risks associated to the authorised GM Plants, has had no influence in the evolution of the EU GMO legislation.

9.6 Conclusions

Risk assessment is a key activity to inform risk managers and help them to properly manage the risk inherent to many of the human activities and ultimately to life on earth. In order to be useful, risk assessment has to be as precise as possible and has to be communicated in a way that can be practically implemented in risk management measures. These measures should be effective in ensuring a tolerable level of risk that is, in itself, highly subjective, and need to be regarded by society as adequate and proportionate. The GMO legal framework built in the EU during the last 25 years intends to solve this difficult equation, finding the equilibrium between the different values at stake, some of which are somehow contradictory. We have tried here to dissect the different factors and values in tension behind the EU GMO regulation and analysed how the changes introduced in the legislation over the last 25 years, have modified these equilibriums. In the last few years, mainly due to the entry into force of the Implementing Regulation 503/2013, the legal framework in which GMO risk assessment is performed has become increasingly prescriptive and rigid, losing the flexibility needed for a case-by-case analysis. This probably makes the system more predictable, but at the same time makes it more difficult to carry out an independent and scientifically sound analysis of each case, which strongly depends on the nature of the GM event and its intended uses. In addition, as the techniques and tools to modify and analyse genomes rapidly evolve, this rigid framework is becoming obsolete. A paradigmatic case illustrating this rigidity, is the difficulty to reach an EU consensus on the status of plants generated by new techniques of genome editing, like the CRISPR/Cas for example. In order to ensure a rigorous science-based risk assessment, there is a need to ensure a legal framework in line with the accumulated experience, flexible enough to allow for a case-by-case analysis of GM plants and products and which can be adapted to respond to the fast evolving science.

Acknowledgements

We thank Howard Davies, Pere Puigdomènech and Jeremy Sweet for helpful discussions and the critical reading of the manuscript.

References

Barton, K.A., Binns, A.N., Matzke, A.J.M. and Chilton, M.D. (1983). Regeneration of intact tobacco plants containing full length copies of genetically engineered T-DNA, and transmission of T-DNA to R1 progeny. Cell 32: 1033-1043.

Brill, W.J. (1985). Safety concerns and genetic engineering in agriculture. Science 227(4685): 381-384.

Commission of the European Communities (1997). Commission decision of 23 July 1997 setting up Scientific Committees in the field of consumer health and food safety. Available at: http://tinyurl.com/z9gkuz8.

Council of the European Communities (1990). Council Directive 90/220/EEC of 23 April 1990 on the deliberate release into the environment of genetically modified organisms. Available at: http://tinyurl.com/hjj6wvr.

Devos, Y., Naegeli, H., Perry, J.N. and Waigmann, E. (2016). 90-day rodent feeding studies on whole GM food/feed. EMBO Reports 17: 942-945.

European Food Safety Authority (EFSA) (2006a). Guidance document for the renewal of authorisations of existing GMO products by the Scientific Panel on Genetically Modified Organisms (GMOs). EFSA Journal 4: 435.

European Food Safety Authority (EFSA) (2006b). Guidance document for the risk assessment of genetically modified microorganisms and their derived products intended for food and feed use by the Scientific Panel on Genetically Modified Organisms (GMO). EFSA Journal 4: 374.

European Food Safety Authority (EFSA) (2007). Guidance document for the risk assessment of genetically modified plants containing stacked transformation events by the Scientific Panel on Genetically Modified Organisms (GMO). EFSA Journal 5.

European Food Safety Authority (EFSA) (2008a). Request from the European Commission related to the safeguard clause invoked by France on maize MON810 according to Article 23 of Directive 2001/18/EC and the emergency measure according to Article 34 of Regulation(EC) No 1829/2003. EFSA Journal 6: 850.

European Food Safety Authority (EFSA) (2008b). Request from the European Commission related to the safeguard clause invoked by Greece on maize MON810 according to Article 23 of Directive 2001/18/EC – Scientific opinion of the Panel on Genetically Modified Organisms. EFSA Journal 6: 757.

European Food Safety Authority (EFSA) (2008c). Request from the European Commission related to the safeguard clause invoked by Hungary on maize MON810 according to Article 23 of Directive 2001/18/EC – Scientific opinion of the Panel on Genetically Modified Organisms. EFSA Journal 6: 756.
European Food Safety Authority (EFSA) (2009a). Request from the European Commission related to the safeguard clause invoked by Austria on maize lines MON863 according to Article 23 of Directive 2001/18/EC. EFSA Journal 7: 1152.
European Food Safety Authority (EFSA) (2009b). Request from the European Commission related to the safeguard clause invoked by Austria on oilseed rape GT73 according to Article 23 of Directive 2001/18/EC. EFSA Journal 7: 1151.
European Food Safety Authority (EFSA) (2009c). Request from the European Commission related to the safeguard clause invoked by Austria on oilseed rape MS8, RF3 and MS8×RF3 according to Article 23 of Directive 2001/18/EC. EFSA Journal 7: 1153.
European Food Safety Authority (EFSA) (2009d). Scientific Opinion on Guidance for the risk assessment of genetically modified plants used for non-food or non-feed purposes. EFSA Journal 7: 1164.
Ewen, S.W.B. and Pusztai, A. (1999). Effect of diets containing genetically modified potatoes expressing Galanthus nivalis lectin on rat small intestine. Lancet 354: 1353-1354.
Fox, J.L. and Norman, C. (1983). Agricultural genetics goes to court. Science 221: 1355.
General Court (2013). Judgment of the general court (seventh chamber) 26 September 2013. Available at: http://tinyurl.com/hp7cc95.
GMO Compass (2008). Maize MON 810: France triggers safeguard clause. Available at: http://tinyurl.com/j98jk48.
Herrera-Estrella, L., Depicker, A., Van Montagu, M., and Schell, J. (1983). Expression of chimaeric genes transferred into plant cells using a Ti-plasmid-derived vector. Nature 303: 209-213.
Joint Working Group on Novel Foods and GMOs (2003). European commission health and consumer protection directorate-general. Guidance document for the risk assessment of genetically modified plants and derived food and feed. Available at: http://tinyurl.com/jq7kb22.
Kolata, G. (1985). How safe are engineered organisms? Science 229(80): 34-35.
Kuiper, H.A., Kok, E.J. and Davies, H.V. (2013). New EU legislation for risk assessment of GM food. No scientific justification for mandatory animal feeding trials. Plant Biotechnology Journal 11: 781-784.
Lee, M. (2008). EU regulation of GMOs. Law and decision making for a new technology. Edward Elgar Publishing, Cheltenham, UK, 274 pp.
Lemaux, P.G. (2008). Genetically engineered plants and foods: a scientist's analysis of the issues, part I. Annual Review of Plant Biology 59: 771-812.
Lynch, D. and Vogel, D. (2001). The regulation of GMOs in Europe and the United States: a case-study of contemporary European regulatory politics. Council of Foreign Relations: 1-39.
Norman, C. (1983). Genetically engineered plants get a green light. Science 222: 35.
Sun, M. (1984). Rifkin broadens challenge in biotech. Science 225: 297.

10. Substituting a fictional 'science' for public accountability: legitimacy problems of the EU's regulatory framework for GM products

L. Levidow
Senior Research Fellow, Faculty of Arts and Social Sciences, The Open University, Walton Hall, Kents Hill, Milton Keynes MK7 6AA, United Kingdom; l.levidow@open.ac.uk

Abstract

Given the EU's continuing conflicts around GM products, how to diagnose the policy problem? Many key actors and commentators have seen the EU's regulatory framework as a failure – but in what sense? Such a judgement has many possible criteria for success – e.g. decisions by the EU regulatory committee to authorise GM products, or clear decisions (qualified majority for or against authorisation) in the EU Council, or public transparency of risk-assessment assumptions, or democratic accountability for agbiotech's contribution to specific agri-development trajectories, or a broader public legitimacy for such societal choices. Those criteria may not be mutually compatible in practice. There is little agreement on what would count as success, much less how to achieve it. Any legislative 'fix' remains elusive. This article analyses the conflict by addressing different questions: What are the basic assumptions of the EU's risk-regulatory regime? How do these generate and structure conflicts over GM products (henceforth agbiotech)? Why did regulatory issues become so contentious? What political interests are at stake? To answer those questions, this article draws on several research projects from 1989-2004 (see Acknowledgements) and a follow-up in subsequent years; the earlier findings can illuminate more recent conflicts. As argued here, the EU's promotional policy for agbiotech has sought legitimacy through a 'science-based' agro-food safety regime, as originally formulated by DG-SANCO, yet expert advice has implicitly incorporated policy commitments favourable to agbiotech. A fictional policy-free 'science' has been substituted for public accountability regarding how GM products are assessed and how they shape the EU's agro-food future. Moreover, agbiotech has been closely bound up with neoliberal agendas. Regardless of legislative changes, therefore, the EU's legitimacy problems around agbiotech will continue.

Keywords: regulatory framework, science-based regulation, sound science, legitimacy, regulatory conflict, precautionary principle, democracy

10.1 EU regulatory framework: policy conflicts

The EU's policy framework has been elaborated and reiterated several times since the 1980s (e.g. CEC, 1993, 2003). In particular the 2002 'strategic vision' for biotechnology emphasised 'science-based regulation', whereby regulatory oversight 'is the expression of societal choices': rules ensure that market mechanisms function effectively through consumer preferences (CEC, 2002). Thus public accountability has been reduced to regulatory oversight of each product, in turn reduced to a fictional 'science' which somehow stands outside policy issues.

DOI 10.3920/978-90-8686-845-2_10, © Wageningen Academic Publishers 2017

In parallel agbiotech regulation has undergone several legislative changes since the original Deliberate Release Directive (EEC, 1990), in particular:

- 2001 revision broadening the scope of potential 'adverse effects' which must be evaluated (EC, 2001; see details below).
- 2015 revision enabling Member States to restrict or prohibit GMO cultivation on their territory, on criteria other than those evaluated in the risk assessment of the European Food Safety Authority (EFSA), i.e. other than grounds of risks to health and the environment (EC, 2015).

Under an extra Commission proposal, member states could flexibly restrict the use of GM food or feed products in their own territory, but (as above) not on grounds which conflict with EFSA's risk assessment (CEC, 2015a). This rule-change was meant partly to facilitate EU-wide approval of GM products, in the knowledge that member states could claim exemptions. But the proposal was overwhelming rejected on feasibility grounds, e.g. because it would obstruct GM protein supplies to Europe, and because the internal market would be disrupted through the border controls necessary to enforce a national ban (EP, 2015).

The defunct proposal was a doubly unfeasible legislative fix for a political problem rooted in the EU's regulatory system. Under the Commission's proposal, as under previous legislative changes, no national restriction on GM products may challenge the EC's official risk assessment, especially by invoking the safeguard clause, which has a contentious history. Under this generic part of EU law, a member may provisionally exempt a product from the internal-market rules on various grounds, including risk to human health or the environment in its territory, as also stated in the EU's first GMO laws:

> Where a Member State has justifiable reasons to consider that a product which has been properly notified and has received written consent under this Directive constitutes a risk to human health or the environment, it may provisionally restrict or prohibit the use and/or sale of that product on its territory (EEC, 1990).

After member states used the safeguard clause against some GM products in the late 1990s, this option was restricted as follows:

> Where a Member State, *as a result of new or additional information made available since the date of the consent* and affecting the environmental risk assessment or reassessment of existing information on the basis of new or additional scientific knowledge, has detailed grounds for considering that a GMO as or in a product which has been properly notified and has received written consent under this Directive constitutes a risk to human health or the environment, that Member State may provisionally restrict or prohibit the use and/or sale of that GMO as or in a product on its territory (EC, 2001).

Thus the clause was limited to contexts where a member state raises new issues or scientific evidence that was not considered by the EU's official risk assessment. To understand why this matters, let us look back at early regulatory conflicts over GM crops, especially at normative issues within risk assessment.

When implementing the 1990 Deliberate Release Directive, by the mid-1990s member states disagreed over how to define 'adverse effects' that must be evaluated and prevented. If the well-known pesticide treadmill were compounded by a genetic treadmill, with the rise of pesticide-resistant crops, then would this be an 'adverse effect'? According to promoters of GM crops, such an outcome would be merely 'agronomic problems', not environmental harm; or such effects would result from agronomic practices, not from the GM crop *per se*. Critics argued that these were false distinctions amongst causes, so that risk assessments should encompass such broader effects (Levidow and Carr, 1996).

Normative issues also arose around potential harm from Bt maize, shortly after the EU gave a marketing consent to this product, meant to protect the crop from the cornborer pest. In the late 1990s new lab evidence indicated that Bt maize could harm non-target insects. In response, the EU's expert advisors argued that any such harm would be less than from agrochemical pesticides, thus presuming intensive agro-industrial methods as the relevant comparator for what counts as 'adverse effects'. Yet in practice, pesticides were rarely sprayed on maize because they cannot easily reach the cornborer. Thus the official experts had invented an extreme baseline, ideologically normalising chemical-intensive methods and conveniently eliminating any uncertainty about any harm greater than from conventional maize.

The new lab evidence of harm was also criticised for test methods not simulating realistic conditions of commercial use. Of course no lab experiment can do so; indeed, the same criticism was warranted even more strongly for many lab tests which showed no harm to non-target insects. According to a meta-study, either the experimental design was inadequate to detect harm, or inadequate information was provided for evaluating the design (Hilbeck *et al.,* 2000). Thus official experts were one-sidedly applying a stringent criterion of 'realistic conditions' to any experiments indicating potential harm, while also maintaining the official pretence of science-based assessment. As a more general pattern, hazard-identification and relevant evidence depend upon normative judgements beyond science, yet these were being either denied or represented as science (Levidow, 2000).

Such policy conflicts led to the 2001 legislative revision broadening the scope of risk assessment. This must include 'any relevant direct, indirect, immediate or delayed effects', as well as any changes in agronomic management practices and their effects, e.g. herbicide sprays. For the commercial stage, moreover, applicants must submit a detailed monitoring plan to identify any adverse effects which were not anticipated in the environmental risk assessment (EC, 2001).

By broadening the range of relevant uncertainties, this legislative change likewise broadened the scientific knowledge necessary for a risk assessment, thus potentially enhancing regulatory legitimacy. Yet expert conflicts continued over how to interpret the broader criteria (Levidow *et al.,* 2005). The new framework structured policy conflict rather than overcoming it. Let us examine the expert conflicts and their implications for recent legislative fixes.

10.2 Expert advice above policy?

The EU's regulatory system has gained its public meaning from changes in the expert advisory system as well as in legislation. After various crises of food safety in the European Union (EU) in the 1990s, regulatory conflicts among member states were impeding EU decisions on agro-food products in several sectors. How were these difficulties understood? The conflicts had arisen from national politics interfering with the proper scientific basis of risk regulation, as well as disagreements among national advisory bodies, according to the problem-diagnosis by DG SANCO (Health and Consumers) and its scientific advisors (SSC, 2003).

As the institutional remedy for such a problem, the EFSA was established in 2002. It was meant to provide expert advice that would be 'independent, objective and transparent'. Its expert advice would help to harmonise 'science-based regulation' and thus facilitate EU political decisions. Institutional reforms were meant to regain public confidence in regulatory decisions. According to the DG SANCO Commissioner, the independence of EFSA '*will ensure that scientific risk assessment work is not swayed by policy or other external considerations*'. Moreover, '*the Authority's reputation for independence and excellence in scientific matters appertaining to food will put an end to competition in such matters among national authorities in the Member States*' (Byrne, 2002).

In such ways, EFSA has been crucial to the Commission's narrative and political authority: regulatory decisions must be 'based on sound science', i.e. 'the principle of a scientific criteria and evidence-based decision' (CEC, 2015b). In practice those criteria have been equated with EFSA's favourable opinions on GM products. This official narrative implies a distinction between unsound/sound science or between non/science, whereby a scientific deficiency amongst national experts explains any disagreement with EFSA's opinions. This strategy puts an enormous burden on 'science' to mediate policy issues. For many risk-assessment issues, scientific ignorance was either denied, e.g. by maligning politically inconvenient results (Levidow, 2002) or else was reduced to calculable uncertainties.

DG SANCO's model sought authoritative expert advice for regulatory harmonisation. In designing EFSA's expert panels, members were selected for their willingness to reach a consensus; this membership criterion minimises the scope for diverse views to be debated or even held (interviews cited in Levidow and Carr, 2007). In EFSA's GMO Panel, moreover, some members were closely involved with the International Life Sciences Institute (ILSI), an industry-funded group originally describing itself as 'a key partner for European industry'. This phrase was later changed to 'a key partner for public health', thus downplaying its partisan character. ILSI claims to create 'neutral fora' for expert discussions, yet it routinely suggests business-friendly approaches to risk-assessment issues. This effort was successful; some EFSA opinions drew on ILSI texts. Eventually these linkages became publicly known, creating a political scandal (CEO, 2012; FoEE, 2006).

All those agendas help explain a general pattern: EFSA's GMO Panel generally gave business-friendly opinions on GM products and generic issues of risk assessment. According to the Panel's early opinions, the available information is adequate for a risk assessment, and the GM product is as safe as its non-GM counterpart. The opinions scrutinised different types of evidence in an asymmetrical way, raising methodological uncertainties about evidence of risk far more than about evidence of safety. Its opinions generally framed scientific uncertainties in such a way that they can be resolved by extra information, or can be readily managed, or can be deemed irrelevant to any risk (Levidow and Carr, 2007).

Meanwhile EFSA made some efforts to accommodate risk concerns from national expert bodies. Yet expert disagreements and public controversy continued, especially regarding risk assessment of GM products. These disagreements have involved normative issues, in particular: which plausible effects would count as 'adverse' (harm), whether some therapeutic antibiotics are dispensable (through pathogens' resistance), what counts as meaningful or adequate evidence for a risk assessment, how to obtain such evidence in a 'cost-effective' way (meaning costs for the applicant), what lab or field experiments should be required, etc. (Levidow and Carr, 2007; Levidow *et al.,* 2005).

Those normative judgements, framing the relevance and burden of scientific evidence, are meant to be explicit. According to EFSA's guidance:

> scientific outputs must be transparent with regard to the data, methods of analysis and assumptions that are used in the risk assessment process... All assumptions should be documented and explained. Where alternative assumptions could reasonably be made, the related uncertainties can be evaluated together with other uncertainties (EFSA Scientific Committee, 2009).

All the normative judgements outlined above, however, remain implicit; they are rarely acknowledged as assumptions, much less as having alternatives. This silence is no accident: by contravening EFSA's own guidelines, risk assessments can be portrayed as a policy-neutral 'sound science', in turn as a politically neutral basis for political decisions.

As a systemic limitation on available evidence, EFSA has largely depended on companies submitting data from their own tests or contractors. The methodologies were often poorly designed to detect potential harm and so were suspected of false negatives; and often the data have not been available for independent checking. This structural weakness has been criticised by regulatory authorities of several EU member states. The problem arises partly by design, given the statutory duties on companies. More fundamentally, it also arises by default: public-sector research institutes have had declining budgets and capacities, themselves becoming more dependent on private-sector finance (Levidow *et al.,* 2002).

Despite all those implicit assumptions and epistemic limitations in risk assessments, the European Commission has generally equated EFSA's favourable opinions with a fictitious 'science'. Within this risk-assessment model, regulatory conflicts have proliferated; new authorisations for GM products are often opposed by a simple majority of member states. As the Commission recently noted, no GM product has gained a qualified majority for or against authorisation in the regulatory committee. Moreover, '*the resistance to GMOs in cultivation has increased in recent years, with many Member States opposing the authorisation of maize 1507 in the Council in February 2014*' (CEC, 2015a). Such disagreements indicate a legitimacy problem around implicit normative judgements in risk assessment.

Analogous regulatory conflicts have arisen in many jurisdictions for a long time, especially in the agbiotech sector. Many scientists deny that GM food has any controversy about health effects, by disparaging or simply ignoring inconvenient test results. Political-economic pressure has been closing down dissent in elite journals, thus disguising corporate interests as 'science' (Krimsky, 2015).

Addressing such regulatory conflicts, the novel concept 'risk-assessment policy' (RAP) was meant to highlight the policy basis for contentious judgements in risk assessment. According to a study of RAPs for various food-safety issues amongst several political jurisdictions, risk assessments always depend on criteria for '*which different types of evidence are included or excluded, as well as the criteria by which the evidence is interpreted*'. Given incomplete scientific knowledge, risk assessments unavoidably give the benefit of the doubt either to public safety or else to commercial interests. Either way, '*some risk-assessment policy issues remain unacknowledged and unaccountable, with policy judgements portrayed as if they were scientific*' (Millstone *et al.*, 2008).

A similar ideological portrayal has continued in the EU system (Millstone, 2015). In a 200-page document, EFSA suggests how risk assessment can better deal with scientific uncertainty, mainly meaning a statistical uncertainty (EFSA, 2015). The document ignores policy and normative judgements such as those above – or else portrays them as science. The EFSA Strategy 2020 involves extensive public consultation, whose scope is limited by the official model of 'science-based' advice and regulation. This elite model derives from EFSA's client relationship with its main stakeholder, DG SANCO (now DG SANTE), which had anticipated or diagnosed regulatory conflict as regulatory competition or political interference in science (Byrne, 2002).

As an alternative model, expert procedures could explicitly acknowledge normative issues pervading risk assessment. Towards such acknowledgement, RAP has been highlighted by international bodies and has been formalised by some national agencies (Millstone *et al.*, 2008). Policy judgements could be made more transparent and publicly accountable by pluralising advisory expertise, encompassing types of counter-expertise (Levidow and Carr, 2007).

Such alternative models have been publicly debated in some places such as France. As a key rationale: when intervening as an expert in a complex domain, scientists necessarily go beyond the limits of their own knowledge; consciously or not, they always function as advocates of a specific cause. This problem-diagnosis motivated a proposal for *expertise contradictoire*, an open space for conflicts among scientific experts (Joly, 2007; Roqueplo, 1996).

Such proposals stimulated a public debate on official expertise in France's GM regulatory procedures. The debate initially led to a more open and transparent regulatory expertise, yet this space was soon constrained. Scientific institutions were protected from scrutiny '*by shifting the boundaries between what counts as science and what counts as non-science*' (Marris *et al.*, 2005). Why? An official space for *expertise contradictoire* would undermine the official pretence of a policy-free 'science', as in the risk-management strategies of the European Commission and member states such as France.

10.3 What societal future through agbiotech?

For regulating agbiotech in the EU, what are the societal stakes? Why did regulatory issues become so contentious? Agbiotech has been deeply embedded in a neoliberal project at many levels – further industrialising agriculture, subordinating public-sector research to agro-industrial interests, and prioritising knowledge for patent claims (Levidow, 2015) – thus deepening regulatory conflicts.

The EU's institutional commitments have linked agbiotech with economic competitiveness, implying that the EU must keep up with technoscientific developments in North America. In EC policy, the entire agro-food industry became discursively 'based on biotechnology', i.e. strategically dependent upon science-led innovation. In particular GM technologies would be essential tools for future growth, competitiveness and environmental improvement (CEC, 1993).

Yet GM crops were initially adopted in US agro-industrial contexts, which became a future model for Europe, where the biotech sector aimed to 'industrialise agriculture', in the words of a lobby group for GM crops (GIBiP, 1990). As key incentives for farmers, EC agricultural subsidies would be reduced and would lose their former link with production. In the view of many biotech company managers, market liberalisation and lower Common Agricultural Policy payments would continue; this trend would be good for future product development, as a greater opportunity to sell GM seeds to farmers (Chataway *et al.*, 2004).

Thus the biotech industry welcomed and promoted greater market pressures for intensifying agri-production methods, as an imperative and opportunity for biotechnological solutions. However, this scenario conflicted with agendas for more extensive agri-methods, 'quality' food, territorial brands, etc. – whose local reputation was threatened by the stigma of GM crops (Levidow and Boschert, 2008). At stake is a choice of rural development models; this societal conflict has taken the form of disputes over product authorisations and 'GM contamination'.

The neoliberal-agbiotech agenda recast research priorities. Priority was given to innovations in molecular genetics, seen as foundational to improvements in efficiency and competitiveness. State-funded research was designated to be 'pre-competitive': to develop basic knowledge that would facilitate downstream competitive innovations. Such policies foresaw the European agri-supply industry becoming integrated into trans-Atlantic multinational companies, contrary to the policy narrative about competition between the US versus EU.

European Community funds for biotechnology research were made dependent upon industry partners committing resources to any project. Research was given a clear economic function, with 'more careful attention to the long-term needs of industry'. According to managers of the DG-Research Biotechnology Division, '*the most vital resource for the competitiveness of the biotechnology industry is the capacity to uncover the mechanisms of biological processes and figure out the blueprint of living matter*' (Magnien and Nettancourt, 1993). This research agenda conceptualised nature as an information machine whose deficiencies had to be corrected, so that such improvement would strengthen European industrial regeneration and competitive advantage.

Under state pressure to become more 'demand sensitive', public-sector research establishments (PSREs) moved towards a more business-oriented organisational culture. As some governments reduced core funding for research, public-sector scientists underwent pressure to seek greater funds from private sources. Moreover, some public funding was shifted from PSREs to universities, which more rapidly expanded into new areas and which more readily utilised short-term funding, e.g. by employing contract researchers. Core funding was reduced or transferred to output financing, i.e. dependent upon competitive bidding for specific projects. In the agro-food sector, these pressures stimulated a shift in priorities from agronomic methods to molecular-level research seeking patentable knowledge. These shifts undermined the public-service roles of PSREs and their public credibility as independent experts, especially when risk controversy arose (Levidow *et al.,* 2002).

As newly prioritised in agricultural research, the pursuit of molecular-genetic knowledge facilitated the extension of proprietary knowledge. In 1988 the Commission proposed a directive granting property rights in 'biotechnological inventions' (CEC, 1988), a key concept from the agri-biotech industry, representing discovery as invention. Opponents raised the slogan, 'No patents on life!' and 'Criminalise biopiracy!' They warned against several harmful consequences: that the Directive would provide an incentive for companies to use GM techniques rather than other methods of improving seeds; that plant patents would deter other forms of seed improvement; and that the mere prospect of litigation could deter other plant breeders from using the germplasm of GM crops.

In the ensuing controversy, 'biopiracy' became a common term for the theft of genetic resources – but with two opposite meanings. For advocates of greater patent rights, 'biopiracy' meant violating the rights of an inventor, by using patented materials without a licence agreement or without paying royalties. For opponents of such rights, 'biopiracy' denoted the patents themselves, on the grounds that biological material should remain freely reproducible as a common resource. After a decade-long debate, proponents gained enactment of the Directive (EC, 1998), but the issues remained contentious. This commercial motive for intellectual property attracted greater suspicion towards the agbiotech sector, even from many scientists who were previously supportive.

For all the above reasons, the EU's conflicts over GM crops have some analogies elsewhere. In the global South, large opposition movements have likewise targeted GM crops as a symbol of neoliberal globalisation and/or of foreign-owned multinational companies. These have been seen as extending farmers' dependence on input-suppliers, e.g. through more concentrated control over on seed varieties, especially through patents. This intense controversy has made regulatory criteria more contentious. (For India, Brazil and Mexico, see Macnaghten *et al.,* 2015: Chapters 2-5; also Macnaghten *et al.,* 2014).

In the USA, whose agriculture was relatively more industrialised than the EU's, GM crops fit the agro-industrial system and were widely seen as reducing its environmental harm, despite grounds for doubt (Krimsky and Wrubel, 1996). Although US NGOs often dispute safety claims for GM products, such debate remains amongst experts only. Wider public dissent has been channelled into non-GM alternatives, especially by protecting and expanding the organic food sector. As a brief flashpoint, the government proposed that organic-certified food could accept GM sources, but this proposal provoked enormous controversy and was soon abandoned. The sharpest political conflicts arise over GM labelling, as proposed in several states by opposed by the Federal government. In all these ways, any systemic conflicts over agri-food futures are channelled into an issue of consumer information and choice.

10.4 Conclusion: beyond regulatory difficulties?

Since the 1990s European citizens have been told that they must accept agbiotech for the EU's economic competitiveness and environmental benefits. The technoscientific development became an instrument of a wider neoliberal agenda linking several aspects – further industrialising agriculture through GM crops, accepting their potential effects as no worse than intensive agri-industrial practices, subordinating public-sector research to agro-industrial interests, and prioritising research agendas for proprietary knowledge.

The 'expression of societal choices' regarding agbiotech was explicitly channelled into regulatory procedures (CEC, 2002), which thereby has born the political burden of societal conflicts around agbiotech. A regulatory harmonisation process, meant to facilitate decisions on GM products, underwent a tension between two political models: One promotes a uniformity to overcome national regulatory divergences, while the other accommodates the diverse national values that generate them (Jasanoff, 2005; Levidow and Carr, 2010).

The European Commission has promoted the former model, harmonisation-as-uniformity, in seeking political authority for decisions. Through EFSA it has sought a fictional policy-free 'science' as the basis for claims that the available scientific information is adequate for a risk assessment that a GM product is as safe as its non-GM counterpart. Through this model, expert advice has regularly incorporated normative or policy judgements, e.g. from ILSI's business-friendly framework, thus downplaying uncertainties about harmful effects as well as their narrow definition. In the name of science, expert advice substitutes for any acknowledgement of RAP judgements or public deliberation on these issues, much less on how GM products relate to the EU's agro-food future.

EU regulatory frameworks involve expert assumptions about and claims for regulatory control over potentially harmful effects, especially by denying any relevant unknowns. Public distrust arises from concerns about this institutionalised culture of science – yet is caricatured as a misperception of 'risk'. This arrangement '*creates not just public opposition to GMOs, but more seriously still, alienation from the institutions supposedly taking responsibility for them*' (Wynne, 2001). In retrospect, such alienation signalled a widespread public resentment against EU institutions usurping national democratic accountability in many policy areas.

The EU's agbiotech regime exemplifies Jean Monnet's legacy for a Europe-wide 'functional integration' through expert committees. Later called the 'Monnet method', expert procedures sought to harmonise regulations in order to achieve a single market across EU member states. Although expert advisors may deliberate policy issues, the procedure tends to depoliticise issues in public arenas, while conveniently offering an expert *fait accompli* to politicians.

This arrangement undermines democracy. A founder of European integration, Altiero Spinelli, warned that technical experts may not take into account 'the broader political context' (Nicolaci, 2012). Or rather, their procedures implicitly incorporate policy judgements, which thereby become less publicly accountable. This exemplifies a wider pattern of policy-based evidence-gathering, contrary to the official stereotype of evidence-based policy (Levidow and Papaioannou, 2015). Indeed, the 'science-based' EU policy framework potentially depoliticised agbiotech.

In this sector, however, the wider harmonisation regime inadvertently made EU policy more vulnerable to attack as democratically unaccountable. A broad opposition alliance branded the regulatory criteria and agbiotech itself as an instrument of neoliberal globalisation (as in many other countries). The official imperative for agbiotech was contested and was turned into a test of the EU's democratic accountability (Levidow and Carr, 2010). The EU's legitimacy problems around agbiotech will continue, regardless of any legislative fix, e.g. past or future changes allowing greater national flexibility in GM product authorisations.

The EU's difficulties in regulating GM crops exemplify wider global conflicts over neoliberal agendas mediated through 'expert-based' regulatory frameworks. Power struggles arise over how to define the issues at stake, the nature of the technology, agri-development choices and the policy basis of scientific evidence. Such regulatory conflicts will persist until the EU creates means for public deliberation of societal choices on agro-food systems: what form of agri-technology for what European future?

Acknowledgements

This article draws on those cited and three EU-wide research projects, all coordinated by the Open University and funded by the European Commission:

- 'GMO Releases: managing uncertainty about biosafety', during 1994-1995
- 'Safety regulation of transgenic crops: completing the internal market?', during 1997-1999.
- 'Precautionary expertise for gm crops', during 2002-2004.
- For the project reports, see http://technology.open.ac.uk/cts/bpg.htm.

References

Byrne, D. (2002). EFSA: excellence, integrity and openness. Speech at inaugural meeting of the EFSA Management Board. EFSA, Parma, Italy.

Commission of the European Communities (CEC) (1988). Proposal for a Council Directive on protection of biotechnological inventions. CEC, Brussels, Belgium.

Commission of the European Communities (CEC) (1993). Growth, competitiveness and employment: the challenges and ways forward into the 21st Century. CEC, Brussels, Belgium.

Commission of the European Communities (CEC) (2002). Towards a strategic vision of biotechnology and life sciences. CEC, Brussels, Belgium.

Commission of the European Communities (CEC) (2003). Towards a strategic vision of biotechnology and life sciences: progress report and future orientations. CEC, Brussels, Belgium.

Commission of the European Communities (CEC) (2015a). Reviewing the decision-making process on genetically modified organisms (GMOs), 22 April. CEC, Brussels, Belgium.

Commission of the European Communities (CEC) (2015b). Speaking points on Regulation on genetically modified food and feed. Council of the EU, Agriculture and Fisheries, 13 July. Available at: http://tinyurl.com/jon2kfg.

Corporate Europe Observatory (CEO) (2012). The International Life Sciences Institute (ILSI), a corporate lobby group. CEO, Brussels, Belgium. Available at: http://tinyurl.com/jpfnvkz.

Chataway, J., Tait, J. and Wield, D. (2004). Understanding company R&D strategies in agro-biotechnology: trajectories and blind spots. Research Policy 33(6-7): 1041-1057.

European Commission (EC) (1998). Directive 98/44/EC of the European Parliament and of the Council on protection of biotechnological inventions. Official Journal of the European Union L 213: 13.

European Commission (EC) (2001). European Parliament and Council Directive 2001/18/EC of 12 March on the deliberate release into the environment of genetically modified organisms and repealing Council Directive 90/220/EEC. Official Journal of the European Union L 106: 17.

European Commission (EC) (2015). Directive 2015/412 of the European Parliament and of the Council of 11 March 2015 amending Directive 2001/18/EC as regards the possibility for the Member States to restrict or prohibit the cultivation of genetically modified organisms (GMOs) in their territory. Available at: http://tinyurl.com/jxn2ek9.

European Economic Community (EEC) (1990). Council Directive 90/220 on the deliberate release to the environment of genetically modified organisms. Official Journal of the European Communities L 117: 15-27.

European Food Safety Authority (EFSA) (2009). Guidance on transparency in the scientific aspects of risk assessment carried out by EFSA, part 2: general principles. EFSA Journal 1051: 1-22.

European Food Safety Authority (EFSA) (2015). Guidance on uncertainty in EFSA scientific assessment. EFSA Journal. Available at: http://tinyurl.com/j4a9xrz.

European Parliament (EP) (2015). Parliament rejects national GMO bans proposal, 28 October. Available at: http://tinyurl.com/h5e72km.

Friends of the Earth Europe (FoEE) (2006). Can we trust EFSA's science? Presentation at meeting with EFSA's GMO Panel. FoEE, Brussels, Belgium. Available at: http://www.foeeurope.org/GMOs.

Green Industry Biotechnology Platform (GIBiP) (1990). The Green Industry Biotechnology Platform, *Agro-Industry Hi-Tech* 1: 55-57.

Hilbeck, A., Meier, M. and Raps, A. (2000). Review on non-target organisms and Bt-plants. Report to Greenpeace International, Amsterdam. EcoStrat GmbH, Zurich, Switzerland. Available at: http://tinyurl.com/jp6nn9e.

Jasanoff, S. (2005). Designs on nature: science and democracy in Europe and the United States. Princeton University Press, Princeton, NJ, USA.

Joly, P.-B. (2007). L'expertise scientifique dans l'espace public: réflexions à partir de l'expérience française. Réalités Industrielles: 23-29. Available at: http://tinyurl.com/gt3knsx.

Krimsky, S. (2015). An illusory consensus behind GMO health assessment. Science, Technology and Human Values 40(6): 883-914.

Krimsky, S. and Wrubel, R. (1996). Agricultural biotechnology and the environment: science, policy and social issues. University of Illinois Press, Chicago, IL, USA.

Levidow, L. (2000). Precautionary uncertainty: regulating GM crops in Europe. Social Studies of Science 31(6): 845-878.

Levidow, L. (2002). Ignorance-based risk assessment? Scientific controversy over GM food safety. Science as Culture 11(1): 61-67.

Levidow, L. (2015). Neoliberal origins of anti-GM protest in Europe. In: Macnaghten, P. and Carro-Ripalda, S. (eds.) Governing agricultural sustainability: global lessons from GM crops. Routledge, London, UK, pp. 183-189.

Levidow, L. and Boschert, K. (2008). Coexistence or contradiction? GM crops versus alternative agricultures in Europe. Geoforum 39: 174-190.

Levidow, L. and Carr, S. (2007). Europeanising advisory expertise: the role of 'independent, objective and transparent' scientific advice in agri-biotech regulation. Environment and Planning C: Government and Politics 25: 880-895.

Levidow, L. and Carr, S. (1996). Special issue: biotechnology risk regulation in Europe. Science and Public Policy 23(3).

Levidow, L., Carr, S. and Wield, D. (2005). EU regulation of agri-biotechnology: precautionary links between science, expertise and policy. Science and Public Policy 32: 261-276.

Levidow, L. and Carr, S. (2010). GM food on trial: testing European democracy. Routledge, London, UK.

Levidow, L. and Papaioannou, T. (2015). Policy-driven, narrative-based evidence-gathering: UK priorities for decarbonisation through biomass. Science and Public Policy 43(1): 46-61.

Levidow, L., Søgaard, V. and Carr, S. (2002). Agricultural PSREs in western Europe: research priorities in conflict. Science and Public Policy 29(4): 287-295.

Macnaghten, P., Carro-Ripalda, S. and Burity, J. (2014). A new approach to governing GM crops: global lessons from the rising powers. Durham University Working Paper, Durham, UK.

Macnaghten, P., Carro-Ripalda, S. and Burity, J. (2015). Researching GM crops in a global context. In: Macnaghten, P. and Carro-Ripalda, S. (eds.) Governing agricultural sustainability: global lessons from GM crops. Routledge, London, UK, pp. 5-33.

Magnien, E. and De Nettancourt, D. (1993). What drives European biotechnological research? In: Blakelely, E. and Willoughby, K. (eds.) Biotechnology review no.1: the management and economic potential of biotechnology. Commission of the European Communities, Brussels, Belgium, pp. 47-48.

Marris, C., Joly, P.-B., Ronda, S. and Bonneuil, C. (2005). How the *French GM controversy led* to the reciprocal emancipation of scientific expertise and policy making. *Science* and *Public Policy* 32(4): 301-308.

Millstone, E. (2015). Letter on EFSA strategy 2020, 18 November. Available at: http://tinyurl.com/j6xywhc.

Millstone, E., Van Zwanenberg, P., Levidow, L., Spök, A., Hirakawa, H. and Matsuo, M. (2008). Risk-assessment policies: differences across jurisdictions. European Commission, Joint Research Centre (DG JRC), Institute for Prospective Technological Studies, Brussels, Belgium. Available at: http://tinyurl.com/yk9tv93.

Nicolaci, F. (2012). Patterns of disintegration: the EU and the emerging European order. European Reform Initiative Press, Brussels, Belgium.

Roqueplo, P. (1996). Entre savoir et decision, l'expertise scientifique. INRA, Paris, France.

Scientific Steering Committee (SSC) (2003). Report on the future of risk assessment in the European Union: second report on the harmonisation of risk assessment procedures. SSC, European Commission, Brussels, Belgium.

Wynne, B. (2001). Creating public alienation: expert cultures of risk and ethics on GMOs. Science as Culture 10(4): 445-481.

11. EFSA's role as an independent authority: a comparative perspective

M. Salvador Martínez
Full Professor of Constitutional Law, Department of Political Law, Faculty of Law, National University of Distant Education (UNED), c/Obispo Trejo 2, 28040 Madrid, Spain; msalvador@der.uned.es

Abstract

The European Food Safety Authority (EFSA) is an EU agency specialised in food safety, an independent authority of a scientific nature intended to be the independent scientific source of recommendation, information and communication of any kind of risk related to the food chain. It began to operate in 2003 and by 2009 was already working at full capacity with all the necessary means in order to fulfil its mandate. In its thirteen years of functioning it has performed very significant work in many different fields (animal health and welfare, biological hazards, pesticides and contaminants, genetically modified organisms (GMOs), nutrition as well as food and feed additives). Its activity embraces some very polemic areas, especially GMOs, and given that the latter are the subject of an intense debate reflecting highly conflicting views, this debate has also inevitably reached the EFSA and its actions as an independent authority. The specific purpose of this work is to analyse the EFSA as an independent authority: what is an independent authority, why is it created, how is its independence legally guaranteed, (organisation, operating procedures, etc.), to what extent does it achieve this objective and what problems does it pose. This analysis will be undertaken with a comparative perspective, given that independent authorities are a category of public bodies that have existed for some time in all the Constitutional States that make up the EU.

Keywords: EFSA, independent authorities, legitimacy, European agencies, risk assessment, democracy

11.1 Introduction

In the 1990s several scandals broke in Europe on the food market: that of the 'mad cows', whose products transmitted Bovine Spongiform Encephalopathy and the contamination of pig and poultry products due to the use of animal feedstuffs with dioxins. These scandals caused considerable alarm amongst the population, seriously damaged the single European market in food and health matters, and questioned the role of the different national agencies responsible for food safety. To address this situation, the European Union passed a Food Safety Regulation, Regulation (EC) 178/2002 (EC, 2002a), which sets the safety requirements that must be met by all food legislation and establishes a risk assessment system based on a specialised technical body, the EFSA.

The EFSA was created as an EU agency specialised in food safety. Among other kinds of independent authorities, it is crucial to underline that the EFSA is an independent authority of a scientific nature. That means it is intended to be the independent scientific source of recommendation, information and communication of any kind of risk related to the food chain. In its thirteen years of functioning the EFSA has performed very significant work in many different fields (animal health and welfare, biological hazards, pesticides and contaminants, GMOs, nutrition as well as food and feed additives).

DOI 10.3920/978-90-8686-845-2_11, © Wageningen Academic Publishers 2017

Its activity embraces some very polemic areas, especially GMOs, and given that the latter are the subject of an intense debate reflecting highly conflicting views, this debate has also inevitably reached the EFSA and its actions as an independent authority.

In this chapter we will analyse the nature and functioning of the EFSA as an independent authority. That includes, first of all a description of what an independent authority is, why it is created and how its independence is legally guaranteed, (organisation, operating procedures, etc.). This will be done by comparing the EFSA with other independent bodies. Secondly, it will be explained to what extent independent authorities achieve their objectives and what problems does their nature pose. This analysis will be undertaken with a comparative perspective, given that independent authorities are a category of public bodies that have existed for some time in all the Constitutional States that make up the EU.

11.2 What is an independent authority? Independent authorities in the Constitutional State

In our geopolitical sphere all societies have been organised politically into Constitutional States designed in accordance with the idea of division of powers: different bodies are created (parliament, government, judges and courts, etc.) according to the various functions to be performed. Well, the idea that specific functions can be correctly fulfilled only by independent bodies has existed since the first Constitutional State. Thus, from the start, the tasks of passing laws and of political direction, which have an essentially political character, are assigned to parliament and government respectively, bodies comprised of political representatives of the citizens; however, the task of resolving conflicts that arise in society applying laws does not have an essentially political, but rather a technical – 'neutral' character, as Montesquieu stated –, and for this reason is not assigned to a political body, but to judges and courts independent of political power, so that they might fulfil their function in a proper and objective manner.

This idea also explains the creation of other 'independent' bodies in the State: Constitutional Courts; certain technical bodies acting as consultants to or collaborators with the administration; and the so-called 'independent authorities or agencies'. The independent authorities are a kind of public body that became widespread in European States at the end of the 20th century and that are characterised by their considerable autonomy with respect to the political bodies of the State when acting in spheres of particular social or economic significance, fulfilling in most cases regulatory functions (which is why they are also known as independent regulatory authorities).

11.2.1 Justification: why is it necessary to create independent authorities?

As I have said, the creation of independent authorities responds to the need for certain functions to be fulfilled outside of politics, and of the partisan conflict and mechanics, to the need to distance certain activities and the taking of certain decisions from essentially political criteria. In other words, there is a desire to remove certain decisions from the sphere of the State in which political parties operate, and in which the political struggle takes place: Parliament, Government and the bodies that depend on them. This argument is a way of responding to the abuse of particracy and political corruption; in fact, independent authorities have always been created, in some cases, given a specific situation of political corruption, and, in others, for preventive purposes, to prevent situations of this nature from arising (Salvador, 2002).

Therefore, the objective is to guarantee a certain 'neutrality' or 'impartiality' in the exercise of a function and to this end this function is attributed to an entity 'removed' ('independent') from the political bodies of the State. The 'independence', therefore, is a degree of qualified autonomy which enables a body to remain sufficiently distanced from political bodies for these not to be able to exercise upon it an influence that prevents it from properly fulfilling its function.

What are these functions that need to be fulfilled independently of the political bodies of the State? The independent authorities that exist in States today were basically created to fulfil two types of functions. Firstly, independent authorities assume functions related to the guarantee of fundamental rights, such as, for instance, the function performed by Data Protection Agencies, which protect the rights of citizens from violation by private companies or by the actions of the State itself. Secondly, in the framework of liberalisation of markets and processes of economic globalisation, functions of an economic or technical nature related to the regulation and supervision of certain sectors of private activity. In this case it is felt that in these economic sectors the traditional mechanisms employed in political decision-making are not appropriate. They are believed to be sectors that are particularly sensitive to political changes and it is necessary to ensure – beyond the vicissitudes of politics – continuity in public policy. It is understood, ultimately, that decisions in these economic sectors should be purely technical decisions, that is, non-political, and should not be influenced by reasons of opportunism or party strategies. This is the argument employed for the creation of independent regulatory authorities.

There are other arguments that have been used alongside that of political neutrality: technical specialisation and efficiency, although these arguments, on their own, are insufficient to justify the creation of an independent body. With regard to technical specialisation, organisms formed by expert specialists in the field exist in the Administrations of all States and fulfil their function properly. For this reason, this argument only makes sense when it is necessary to guarantee both technical specialisation and political neutrality; moreover, what this argument demands is that the independent authority to be created be formed by technical experts in the field. The same thing occurs with the argument of efficiency: only upon recognition of the need for political neutrality does the efficiency argument make sense, as a criterion to determine the regime of the independent authority that is going to be created, and more specifically, so that the independent authorities are armed with more flexible and faster (more efficient) procedures compared with the traditional procedures of the Administration, which are on occasions thought to be excessively rigid and slow (Salvador, 2002).

11.2.2 Functions: why the independence?

Most independent authorities are assigned the so-called regulatory function. This is not easily defined because there is no single model of regulation, since the latter can pursue different objectives and do so in different ways. It could be said, however, generally speaking, that the function of regulation consists in regulating in a broad sense, in other words, in ordering and supervising all private activity in a sector to ensure its proper functioning. To this end, the authority would embrace all those activities (legislative, executive and parajudiciary) that were necessary depending on the specific needs of each sector. For example, the North American regulatory authorities (the independent agencies) establish how private companies must act in the regulated sector, fix and control prices, approve technical and quality standards, regulate access to an economic activity, control compliance with laws and intervene in the event of non-compliance, may grant certain subjects special status, and impose economic sanctions, establishing the quantity themselves. In Europe the task assigned to regulatory authorities is more limited: it consists above all in supervising the functioning of a sector and not so much, or at least not to the same degree as in the USA, in regulating in the strict sense. The latter is a task that the European regulatory authorities fundamentally share with the Government. Furthermore, many European regulatory authorities act only as advisory or support bodies: they have been designed as bodies of experts in a specific field, which is why they issue mandatory and/or binding reports, perform studies, and advise, when this is required of them.

11.2.3 Organisation: how to guarantee independence?

The independent authorities have been created without a previous plan, without a legal regulation common to all of them. For this reason they do not all have the same form of organisation. There exist, however, important coincidences with regard to the mechanisms that have been adopted to distance these bodies from the political organs of the State, which could be ordered in the following manner:

1. In order to guarantee organic autonomy these authorities are created by law, and they are attributed a separate legal personality from the administration.
2. To guarantee functional autonomy it is essential that they cannot receive any kind of order or instruction from other bodies; it is also important to ascribe them exclusively the function for which they were created, and, if they share the exercise of that function with other bodies, it is fundamental that there exist sufficiently clear demarcation of the scope of action of each.
3. To guarantee the personal autonomy of the members of these institutions it is vital to limit the participation of the Government in the election and removal of its members. For this reason, their election usually involves the intervention of the government, Parliament, other public entities, and, even, private associations and citizens; the systems of election are very varied. With regard to the removal of members, discretional removal by the Government is prohibited: members can only be removed in the cases established by law (incompatibility or supervening incapacity, conviction for intentional crime and serious breach of obligations). It is also fundamental to establish eligibility requirements to ensure that members are qualified persons, and reasons of incompatibility that prevent members from holding another public position, profession or employment related to the function of authority. Finally, there is an attempt to prevent the term of office from coinciding with that of the government and, moreover, in some cases renewal is effected in parts, in stages, and re-election is forbidden or only one renewal permitted.
4. In order to guarantee financial autonomy the aim is for these authorities to have a source of self-financing, such as the collection of fees, and they are granted administrative and staffing autonomy.
5. Finally, to avoid political control of these authorities, there is the utmost limitation of the political control that may be exercised by Government and Parliament. But this limitation of purely political control (of criteria of political opportunity) must be compensated for with a strengthening of non-political means of control (legal, financial, etc.). Thus the following are required: absolute transparency in the composition and functioning of the authority; that all actions be paid public; decision-making processes which guarantee objectivity and motivation;

and passing of annual or pluriannual activity plans and accountability via the presentation of periodical or annual reports.

11.2.4 Assessment: the level of independence achieved and its cost

From the experience of the years of operation of independent authorities in constitutional States it can be said that the means adopted to guarantee independence have formally achieved their objective and that on a day-to-day basis these authorities act with reasonable autonomy, although there are two risks of which one needs to be aware.

Firstly, given that most appointments of members of independent authorities are the responsibility of Government and Parliament, the latter may nominate persons who, fulfilling the necessary requirements, are 'close' to political parties and may be tempted to seek to influence them in those cases in which they have a special interest. This danger is apparent to a different degree between one country and another, depending on the characteristics and dynamics of the party system, and whether in the design of the organisation of the State there exists sufficient checks and balances that make it possible to limit the natural tendency of parties to create patronage networks. In order to address this danger it is vital to create a suitable political culture in all actors: in citizens, in political parties, and especially in the authority itself, which has to be aware of the need to gain the reputation of 'independent' and of the need to defend that independence once it is achieved (Cushman, 1941).

The second danger is the capture by the regulated sector, in other words, the influence that may be exerted upon the authority by companies within the sector and the risk of excessive identification of the authority with the sector's interests. This is a phenomenon that has occurred particularly in the USA (Quirk, 1981), but not in Europe, above all because here many of the regulatory powers are in the hands of Governments.

Meanwhile, the independence of these authorities has had certain costs, especially for the democratic principle. Independence has been achieved by equipping these authorities with a legal system in which they are distanced from political parties, from the political organs of the State and, therefore, from the channels via which the accountability mechanisms of political responsibility and political control act in the democratic State, mechanisms without which there is no democratic legitimacy. This is why independent authorities are said to lack democratic legitimacy. There has been an attempt to compensate for this deficit fundamentally by appealing to 'technocratic legitimacy': the activity of independent authorities is legitimised because the people who form them are specialists in the field who take all decisions objectively solely on the basis of technical criteria. However, this appeal is valid only if the authority only takes technical decisions and leaves the political decisions to be adopted consequently to political bodies, as occurs with authorities that are essentially advisory. This is not the case of those authorities that have broad regulatory powers, because in these cases the decisions would not be purely technical but also political (Shapiro, 1996). Ultimately, as has been said, the democratic deficit of these authorities, which is the cost to be paid for their political independence, may be overcome via strong connection to the law and the strengthening of other controls (legal, technical) that can compensate for the limitation of political controls.

11.3 Independent authorities in the European Union

11.3.1 Some introductory notes

In the European Union there are also independent authorities, such as the Agency for the Cooperation of Energy Regulators, the European Agency for Safety and Health at Work or the European Railway Agency. These authorities respond to the same idea of the independent authorities of constitutional States, and the way they are organised is very similar, although they act in a different context. This is because the European Union, as we know, is not a constitutional State, but rather an international organisation of States immersed within a process of increasing integration.

The organisation of the EU does not respond to the principle of the separation of powers in the same way of national States and for this reason European institutions are not comparable to the legislative (Parliament) and executive (Government and administration) powers of States, but all share to a greater or lesser degree, and depending on the case, the functions of political direction, passing of laws, execution and control. The European Parliament is the body directly chosen by citizens and the European Council is formed by the Heads of State or of Government of the Member States. The EU Council or Consilium, meanwhile, is comprised of a minister representing each Member State, and added to these is the Commission, which is the body that represents the general interest of the Unión. The latter is formed as a sort of Government of the EU, and therefore is not composed of representatives of Member States, but of a President and a group of commissioners elected by other bodies, and has broader powers than the other institutions.

In this institutional context, the first independent European authority was created in 1975 and today there are more than 25. Most were created from 2002 onwards, when the EU needed a larger administration to tackle all the tasks it faced. That was when the decision was made to create an administrative model acceptable to all the States and that would enable the EU to grow. That model is formed of agencies, bodies with their own legal personality, that are fundamentally divided into two groups (EC, 2002b): the executive agencies, created by the Commission, for a specific period of time, which help the Commission to run EU programmes, are retain close links with it; and the independent authorities, independent bodies created to perform technical and scientific tasks that help the EU institutions implement policies and take decisions, also support collaboration between the EU and national governments, and are set up for an indefinite period.

11.3.2 Justification and functions of the European independent authorities

The creation of independent authorities in the EU responds to similar motives as in national States. Firstly, it reflected the need to separate certain tasks from the Commission, the EU government body. In the late 1990s the mad cow crisis, the crisis that affected the Santer Commission (accused of fraud and mismanagement) and other similar episodes led to a crisis of confidence in the existing management model which directly depended upon the Commission; furthermore, the Member States, and the other Community institutions, did not want the Commission to become stronger and preferred a form of administration more independent of the latter in which the States could participate. Secondly, the idea was to relieve the Commission of purely technical tasks and assign these to technically specialised bodies. Thirdly, the aim was to create bodies that might act swiftly and efficiently, but that were also smaller and more transparent than the committees and departments that depended on the Commission and whose lack of transparency was highly criticised. And, finally, the intention was to promote the participation of Member States, anticipating that the latter would take part in the appointment of the members of these authorities (Tarrant and Kelemen, 2007; Vírgala, 2011).

Given that the European sphere differs from the national one, the objective of the promotion of the independence of decentralised agencies has been to equip them with a qualified autonomy that enables them to fulfil their technical function; in other words, an autonomy which keeps them removed not only from Commission, which is the institution that most clearly seeks to expand its powers, but also from the other EU institutions, from the Member States (which principally look after their own national interests) and from the operators in the sector affected (EC, 2005)

The function assigned to these agencies is that of helping the EU to elaborate and apply its policies. This means that these agencies basically perform tasks of study and counselling; in general they do not have decision-making powers, although there are differences between those that are purely advisory bodies, those that can predetermine with their reports the decisions taken by the Commission and those that have minor decision-making powers for individual cases. This is due to the Meroni doctrine established by the EU Court of Justice, according to which the decision-making powers of the European institutions are an exercise of sovereign powers ceded by States to institutions via treaties and cannot be subdelegated to agencies (Vírgala, 2011). In spite of this, the agencies have considerable relevance because their technical opinion may be decisive in the decision-making process in the EU, and because they constitute an essential element for the coordination of national and European administrative systems.

11.3.3 Organisation of the European independent authorities

With regard to the organisation and legal framework of these agencies, there are many differences between them, which are reflected in very different degrees of formal independence (Wonka and Rittberger, 2009, 2010). For this reason in July 2012 the EU adopted a *Joint Statement and Common Approach of the European Parliament, the Council of the EU and the European Commission on decentralised agencies* (European Parliament, Council and Commission, 2012), which established a model of organisation and control which all agencies should aim to follow. In December 2012 the *Roadmap on the follow-up to the Common Approach on EU decentralised agencies* (EC, 2012) was passed, with the action plan to achieve alignment with this model, especially with regard to attaining more balanced governance, improved efficiency and accountability and greater coherence.

The measures established by these documents to guarantee the independence of these agencies are very similar to those of national independent authorities and are aimed at ensuring their organic autonomy (creation by Regulation, separate legal personality), functional autonomy (independence when carrying out their work), personal autonomy (limiting appointments of a political nature, demanding a specific qualification, avoiding conflicts of interest, etc.), and financial autonomy (managerial autonomy, staffing autonomy, certain own revenues, etc.). There is also limitation of the purely political controls to which they are subjected, while other controls are reinforced (transparency, accountability, etc.). In spite of this there are still significant differences between the organisation of different agencies, and the level of formal independence continues to vary.

With respect to the degree of material independence, studies carried out to date (Egeberg and Trondal, 2010, 2016) reveal that the actor with most influence upon the agencies is the Commission. Although its representatives are minority on the Management Boards of the agencies, these Boards meet infrequently, tend to be inefficient (many members, many languages, etc.) and the Commission representatives are able to exert political influence on them because they are very familiar with the issues and are particularly attentive to the actions of the agency. Furthermore almost all the work performed by the agency is for the Commission and the agency tends to depend on the latter more than on other institutions. The studies also show the importance of each agency adopting a clear attitude when it comes to earning recognition as an independent agency and really becoming an independent and technical reference in the sector. This attitude, which goes beyond what might be established by laws, is the best guarantee of the material independence of the agency and also serves to legitimate its actions. In this regard there have also been studies that show the differences that exist between agencies, and even within the same agency, from one period to another (Groenleer, 2009).

11.4 The EFSA as independent authority

The EFSA is one of the decentralised agencies in the EU. As has been said, it was created in 2002, by Regulation (EC) 178/2002 (EC, 2002a), to become the independent scientific authority in the sector of food safety.

The aforementioned Regulation expresses a clear and unambiguous will to create an independent agency. Its recitals explain that the free movement of safe and wholesome food is an essential aspect of the internal market, and food safety and the protection of consumer's interests is of increasing concern to the general public, non-governmental organisations, professional associations, international trading partners and trade organisations. For this reason, the EU has chosen a high level of health protection as appropriate in the development of food law and has demanded that all laws and decisions in this field be based upon analysis of health risks. The EU is aware that it is vital that there exists confidence that the legislation and the decisions have a scientific and technical basis, and that the scientific and technical issues in relation to food and feed safety are becoming increasingly important and complex. All this makes it necessary for the EU to have high-quality, independent and efficient scientific and technical support. The previous system, which depended upon the State health systems, was no longer considered to be able to respond to increasing demands on it. For this reason the decision was taken to create the EFSA, as an independent scientific reference body.

Thus, the EFSA responds to the prototype of independent authority of a technical-scientific nature which, as I have said, is created to guarantee that technical decisions are taken on the basis of scientific criteria and without the interference of political criteria. Its task is to serve as an advisory or reference body in technical questions, so that other institutions may, basing themselves on its reports, take the appropriate political or legislative decisions. In this sense, and given the essentially technical nature of the issues addressed by the EFSA, there can be little objection to the motive that justifies its creation as an independent authority. More debatable might be the case of a regulatory agency in social or economic areas, because the social sciences are of a different scientific nature to the natural sciences, less objective and more appreciatory in some aspects, and it could be argued that decisions in this field are not exclusively technical. This is a criticism that has been levelled on many occasions regarding this type of agency.

11.4.1 Functions and powers (and functional autonomy)

The main function of the EFSA is to provide the Community institutions and the Member States with scientific advice in all cases provided for by Community legislation and with regard to any question included within its remit. The EFSA has a broad remit, covering the entire food chain from farm to fork with topics related to animal health and welfare, biological hazards, pesticides and contaminants, GMOs, nutrition as well as food and feed additives. Its function is also to promote uniformity between national and community policies in matters of food health, and collaboration between Member States (their relevant agencies), international bodies and other subjects involved (arts. 22 and 23 Regulation (EC) 178/2002 (EC, 2002a)).

In order to perform these functions the EFSA does not have regulatory, executive or parajudicial powers. The EFSA only has authority to issue Scientific Opinion, Statement, Guidance Document or another type of output. It does not have decision-making powers, but its opinions, given that these are very technical questions, have considerable value and influence on the decisions taken on the basis of these opinions. For this reason it has been described as a 'pre-decisive 'agency. The scientific advice of the EFSA should express the scientific answer to the question that has been posed, depending on the state of scientific knowledge and of science at that moment and in that field. This is the EFSA's obligation. On the basis of the scientific advice of the EFSA the political bodies will be responsible for taking the appropriate decision. The EFSA's advice should be taken into account (art. 3 Regulation (EC) 178/2002 (EC, 2002a) with regard to the elaboration of food legislation, for example), but the political bodies have a relatively wide margin of decision, depending on the field and the case, and may even deviate from the agency's recommendation. If they do so, they must justify their decision and that justification must be sufficient and reasonable, because this is how there is legal control against arbitrary decisions; in the event of justification being considered insufficient, the decision in question may be annulled (Craig, 2012).

Thus for example, in the field of GMOs, no genetically modified food or animal feed may be commercialised without specific authorisation (Directive 2001/18/EC (EC, 2001) and Regulation (EC) n° 1829/2003 (EC, 2003)). For this, the first step is the scientific evaluation performed by the EFSA, which is responsible for carrying out risk analysis of the food or feed in question. Once the EFSA has issued its verdict, it will be the Commission and/or the Member States who will decide, on the basis of the EFSA report, whether or not to grant the authorisation. If the EFSA considers that a crop or food is not safe, that is final. But if it considers it to be authorisable (that it meets the risk standards) this does not automatically translate into authorisation.

The political decision-making bodies may depart from the proposal of authorisation made by the EFSA and decide against authorisation, because these organs enjoy a degree of discretion that takes into account criteria of a socio-economic and political nature. Addressing the more problematic aspects of the authorisation of the use of GMOs is not, therefore, the task of the EFSA but corresponds to the Commission and to the bodies in which States are represented. Thus, arts. 7 and 19 of Regulation (EC) n° 1829/2003 (EC, 2003) establish that:

> After receiving the opinion of the Authority, the Commission shall submit a draft of the decision to be taken in respect of the application, taking into account the opinion of the Authority, any relevant provisions of Community law and other legitimate factors relevant to the matter under consideration. Where the draft decision is not in accordance with the opinion of the Authority, the Commission shall provide an explanation for the differences.

Having defined the functions and powers of the EFSA it should also be noted that the degree of functional autonomy guaranteed to it by law is high: the agency's area of activity is clearly defined, this area is delimited vis-à-vis the sphere of activity of other agencies and institutions, there is affirmation of the EFSA's independence in the exercise of its technical function, and all its members are required to undertake to act with independence and in the public interest, independently of any outside influence (art. 37.1 and 2 Regulation (EC) 178/2002 (EC, 2002a)).

11.4.2 Organisation (organic, personal and financial autonomy)

The organisation of the EFSA is fundamentally established in Regulation (EC) 178/2002 (EC, 2002a), although measures established in other European legislation on agencies are also applied. This agency has the organic autonomy that defines independent authorities: it was created by a Parliament and Council Regulation, and has its own legal personality (art. 46.1 Regulation (EC) 178/2002 (EC, 2002a)).

In order to determine the degree of personal autonomy one needs to analyse the system of election and removal of members, the conditions they must fulfil and the measures taken to avoid conflicts of interest.

Appointments are the responsibility of the European institutions, many of them following an open and competitive process; there can be no discretional removal; and all members of the agency are required to be specifically and broadly qualified. Let us look at this in each organ of the EFSA. The EFSA is composed of a Management Board, an Executive Director, an Advisory Forum, a Scientific Committee and 10 Scientific Panels (arts. 25 a 28 Regulation (EC) 178/2002 (EC, 2002a)).

Management Board, Executive Director and Advisory Forum of the EFSA

1. The Management Board is comprised of 15 members, which is a governable quantity, compared with the Boards of other agencies that are rendered far more inefficient by their high number of members. Of these 15, one is a representative of the Commission and the other 14 are nominated by the Council, in consultation with the European Parliament, from a list prepared by the Commission containing a considerably larger number of candidates than those to be appointed. In other words, in the selection of the members the Commission has greater influence than the other institutions, though all participate, and there are no members directly elected by States, as there are in other agencies. Its members must meet the highest standards of competence and have

a broad range of relevant expertise; furthermore, four of the members shall have their background in organisations representing consumers and other interests in the food chain. The members' term of office shall be for four years, renewable only once. The Board is the central organ of the agency and is responsible for establishing the agency's general policies and work plan, and for nominating the members of the other organs. Decisions must be taken after a debate and vote. The Executive Director and the Chair of the Scientific Committee take part in meetings, but without voting rights.

2. The Executive Director is appointed by the Management Board on the basis of a list of candidates proposed by the Commission after an open competition. The term of office is five years and is renewable. The executive director is the executive organ of the agency and its legal representative. He/she is responsible for the day-to-day administration of the Authority and for the implementation of work programmes and the decisions adopted by the Management Board.
3. The Advisory Forum is composed of representatives from competent bodies in each of the Member States which undertake tasks similar to those of the Authority. To compensate for the fact that the agency Management Board does not include representatives of the Member States, as is the case in other agencies, this Forum was created. It is chaired by the Executive Director and meets no less than four times a year, at the invitation of the Chair or at the request of at least a third of its members. Its task is to advise the Executive Director in the performance of his/her duties in particular in drawing up a proposal for the Authority's work programme. The Advisory Forum is the venue for cooperation between Member States and the Agency, for which purpose it is assigned special powers.

The Scientific Committee and Permanent Scientific Panels

The Scientific Committee and Permanent Scientific Panels are the agency's working bodies, responsible for providing scientific opinions, each within their own spheres of competence. The members of the Scientific Committee and of the Scientific Panels are selected via an open process of competition open to all candidates wishing to apply and fulfilling the necessary qualifying requirements. Three external experts, chosen for their Independence and scientific expertise, ensure the fairness of the selection process. After ensuring that the candidates selected have no conflicts of interests, the Advisory Forum is consulted and, finally, they are appointed by the Management Board, upon a proposal from the Executive Director, for a three-year period that is renewable. Their CVs and declarations of interest are published on the EFSA webpage.

The Scientific Committee is composed of 16 members: the Chairs of the 10 Scientific panels and six independent scientific experts who do not belong to any of them. Scientific Committee Members must be prominent scientists from across Europe, of recognised scientific excellence with competences spreading across disciplines within the EFSA's remit, seniority and prior experience with scientific bodies. The Scientific Committee is responsible for the general coordination necessary to ensure the consistency of the scientific opinion procedure (working procedures and harmonisation of working methods).

Meanwhile, the Scientific panels are composed of 21 members, independent scientific experts. There are currently 10, each of them devoted to a different area within the food chain: food additives, plant protection products, biological hazards, contaminants in the food chain, plant health, etc. One of them is devoted to GMOs. Specifically, the GMO Panel Members are scientists from across Europe with expertise in: food and feed safety assessment; food and genetic toxicology, immunology, food allergy, human and animal nutrition, dietary exposure, biochemistry and metabolism, food chemistry, compositional analysis statistics and field trial design, animal feeding trials; environmental risk assessment; and molecular characterisation and plant science.

Management of conflicts of interest within the EFSA and financial independence

With regard to conflicts of interest, the legislation requires a declaration of interests in order to avoid situations of incompatibility or conflict of interests with the Agency (art. 37 Regulation (EC) 178/2002 (EC, 2002a)) and the EFSA has developed this extensively approving the Policy on Independence and Scientific Decision-Making Processes (EFSA, 2011b) and the Decision of the Executive Director on Declarations of Interests (EFSA, 2014). All members of EFSA's Scientific Committee, Scientific Panels, and Working Groups and by all other EFSA experts as well as members of the Authority's Management Board and Advisory Forum are obliged to make annual declarations of interest that are compiled in a public data base. The Declaration of Interests contains details of current activities and those completed in the last five years in a wide variety of categories: ownership or other investments, including shares; research funding; intellectual property rights; interests of close family members, etc. All EFSA experts are also asked to declare any specific interests which might be considered prejudicial in relation to items on the agenda of the meetings they attend. If a conflict of interest is identified, it is recorded in the minutes and EFSA takes the measures specified in the procedure for handling conflicts of interest.

Finally, the means of financial independence are typical of this type of body (arts. 43 and 44 Regulation (EC) 178/2002 (EC, 2002a) and Commission Delegated Regulation 1271/2013 (EC, 2013)). The revenues of the Authority shall consist of a contribution from the Community and, from any State with which the Community has concluded agreements, and charges corresponding to the Agency's activities. The draft budget is proposed by the Agency (drawn up by the Executive Director and approved by the Management Board), which forwards it to the Commission, along with a provisional list of posts and the work programmes. The Commission will include it in the EU budgets and it will be authorised along with the latter. The authority has management autonomy, which corresponds to the Executive Director.

11.4.3 Functioning of the EFSA

The EFSA organises its work programme – annual and multi-annual – according to priorities agreed with the European Commission and other partners. The working procedure is as follows. The workflow for the EFSA's scientific outputs runs from the moment the EFSA receives a request for scientific advice or initiates its own activity to the moment it publishes its findings. Normally the EFSA acts to respond to a request, but it may also act upon its own initiative to examine emerging issues and new hazards and to update its assessment methods and approaches.

When the EFSA receives requests for scientific advice mainly from the European Commission, EFSA considers its contents and discusses it with the Commission. Following these discussions, EFSA and the Commission agree on a mandate, which includes the final terms of reference and a mutually agreed deadline. The mandate is then assigned to one of EFSA's Scientific Panels or its Scientific Committee. On the EFSA webpage there is a Register of Questions where it is possible to consult all the requests accepted by the EFSA and their current status.

EFSA usually sets up a Working Group of experts to carry out the risk assessment. The working group – typically comprising members of the Panel plus additional scientists from specialist fields – assesses the scientific information available, which may include data supplied by Member States, research institutes or companies. If there is a need for further data it may draw on EFSA's data collection networks or launch an open call for data on the EFSA website. The working group develops a draft and submits it to the Panel for discussion. The assessment is adopted by a majority of Panel members – with any minority opinion recorded – at a plenary meeting of the relevant Panel. If the Panel does not adopt, the document is sent back to the working group for further examination and to prepare a revised draft for adoption at a following plenary meeting. The output is then published on the EFSA website in the EFSA Journal, our open-access, online scientific journal.

Given that the EFSA performs a task of scientific research, one expects it to act in accordance with the scientific standards established by the scientific community at a particular time, and all necessary measures must be adopted in order for the operating procedure to meet these standards. In this sense it is worth highlighting the measures adopted by the EFSA itself, which has developed a comprehensive body of good risk assessment practices and methodologies to guide the work of its Scientific Committee, Scientific Panels and its scientific staff, to ensure their opinions respect the highest scientific standards (EFSA, 2011b). Fundamental too is collaboration with other EU Member States, national authorities, EU institutions, scientific organisations, civil society stakeholders (consumer organisations, non-governmental organisations, industry associations, media), international organisations and other countries, and this is provided for by Regulation (EC) 178/2002 (EC, 2002a) for numerous instances (art. 33, for example, for data collection).

11.4.4 Controls, transparency and accountability

With regard to controls, what defines independent authorities is the limitation of political controls and the strengthening of other forms of control. The EFSA system of controls corresponds to this model.

Firstly, the EFSA is subject to the general legal controls that exist in the EU and to other ordinary controls, such as those represented by the European Ombudsman and the European Anti-fraud Office. Financial control is subject to the general system and corresponds to the EU Court of Auditors (arts. 43 and 44 Regulation (EC) 178/2002 (EC, 2002a) and Commission Delegated Regulation (EU) No 1271/2013 (EC, 2013)).

Secondly, in the *Joint Statement and Common Approach of the European Parliament*, (European Parliament, Council and Commission, 2012) it was agreed to promote measures to reinforce responsibility and control of the agencies. In general it was a case of strengthening accountability mechanisms that already existed, but there was also the introduction of the so-called Warning System. This system enables the Commission to react when it believes that the management board of an agency is going to take decisions that might violate the agency's mandate, breach EU legislation or openly contradict the political objectives of the EU. In these cases, the Commission can ask the Management Board to refrain from taking that decision. If it fails to do so, the Commission will officially inform the European Parliament and the Council, so that the three institutions can react quickly. This is a similar measure to those that exist in constitutional States in order to react to extraordinary situations and there is normally no need to use it. In the EU to date it has not been activated in any agency (EC, 2015).

Thirdly, Regulation (EC) 178/2002 (EC, 2002a) which regulates the EFSA contains a very complete control model which includes the requirement of transparency, accountability and additional technical controls:

1. The EFSA is subject to a strict regime of publicity and transparency, which is the first element necessary in satisfactory control system (art. 38). All its activities are carried out with a high degree of transparency. In particular it must make public: agendas and minutes of the Scientific Committee and the Scientific Panels; the opinions of the Scientific Committee and the Scientific Panels; the information on which its opinions are based; the annual declarations of interest made by members; the results of its scientific studies; the annual report of its activities; requests from the European Parliament, the Commission or a Member State for scientific opinions which have been refused or modified and the justifications for the refusal or modification. The Management Board shall hold its meetings in public, on a proposal from the Executive Director, unless it decides otherwise for specific administrative points of its agenda, and may authorise consumer representatives or other interested parties to observe the proceedings of some of the Authority's activities. It is also legally established that the agency must plan its activity proposing work plans, which must be presented to the European Parliament, to the Council, to the Commission and to the Member States (art. 26.3) And upon completion of the period of execution of the work plan the agency must present the general report on activities, with all the tasks performed, to the European Parliament, the Council, the Commission, the Court of Auditors, the Economic and Social Committee and the Committee of the Regions (art. 26.3).
2. It is worth highlighting, finally, that a new technical control has been introduced. Every six years the EFSA, in collaboration with the Commission, will request an independent external evaluation of its achievements on the basis of the terms of reference issued by the Management Board, which will assess the working practices and the impact of the Authority, and will take into account the views of the stakeholders, at both Community and national level. The Management Board of the Authority shall examine the conclusions of the evaluation and issue to the Commission such recommendations as may be necessary (art. 61).

11.5 Concluding remarks

From the legal-technical point of view, the EFSA responds perfectly to the model of independent authority. As we have seen, the motives given for its creation as an 'independent' agency are completely reasonable: in the food safety sector decisions should be taken on scientific grounds. The elaboration of those scientific grounds is assigned to a technical agency, scientifically qualified, and independent of political interference or private interests. The political bodies will be responsible for taking decisions on the basis of these scientific grounds.

Its existence as an independent authority justified, the legal system with which it has been equipped is appropriate and the measures adopted are designed to ensure sufficient organic, personal, functional and financial autonomy (albeit improvable in some instances). But let us not forget that this formal independence is always relative, it cannot be absolute, in the sense that the EFSA, and all the other independent authorities, exist and function within a larger organisation (be it a State or the EU), in which they are integrated. The independence attributed to them is the means of qualified autonomy they require in order correctly to fulfil their function. Moreover, beyond the formal independence established by laws, it is the agency itself that must earn the label of independent, acting as such and reacting to any attempts to influence it.

It was pointed out at the beginning of this work that independence comes at a cost: the limitation of political control and, therefore, of political legitimacy. However, this cost is not the same in every case. The independent regulatory agencies that exist in some States with broad decision-making powers and capacity to shape public policy in the regulated sector do have lack of democratic legitimacy, because they are participating in political decision-making and in a democratic State one has to be legitimised to do this. The EFSA, however, is the prototype of independent advisory agency, given that it is responsible only for the technical decision while it is the political body (legitimised to do so) that adopts the political decision. Its influence may be decisive, but it does not make the final decision, which frees it from criticism for lack of democratic legitimacy.

The regulatory framework established for GMOs in Directive 2001/18/EC (EC, 2001) (partially modified by Directive 412/2015) (EU, 2015) is a good example of this. This legislation entrusts to the EFSA an independent scientific assessment that is, in any case, a *sine qua non*, and leaves in the hands of the political decision-making bodies the appreciation of socio-political and economic criteria, of opportunism even, which by definition cannot correspond to the EFSA

Essentially, Directive 412/2015 divides into two phases the process of political decision-making with regard to GMOs. Firstly, a decision is taken at Community level to authorise, or not, the use of a GMO and, subsequently, States are permitted, with regard to crop authorisation, partially or totally to restrict the agricultural use of said GMO in their territory. In both cases these are decisions taken after the risk assessment commissioned to the EFSA and which must be based upon different criteria from those that are exclusively the competence of the EFSA. This means that if a State or body within the EU, such as actors in the sector or citizens, were in possession of relevant information according to which it would be appropriate to review risk assessment performed by the EFSA (for example, following the abstention of new scientific data), the phases of political decision-making would not be where or when to review the EFSA decision: there exists in accordance with Regulation (EC) 178/2002 (EC, 2002a) a mechanism that activates review by an independent agency.

References

Craig, P. (2012). EU administrative law. Oxford University Press, New York, NY, USA.

Cushman, R. (1941). The independent regulatory commissions. Oxford University Press, New York, NY, USA.

Egeberg, M. and Trondal, J. (2010). EU-level agencies: new executive centre formation or vehicles for national control? ARENA, University of Oslo, Oslo, Norway. Available at: http://tinyurl.com/z9x6b47.

Egeberg, M. and Trondal, J. (2016). Agencification of the European administration: connecting the dots. ARENA, University of Oslo, Oslo, Norway. Available at: http://tinyurl.com/jq2hyj3.

European Commission (2001). Directive 2001/18/EC of the European Parliament and of the Council of 12 March 2001 on the deliberate release into the environment of genetically modified organisms and repealing Council Directive 90/220/EEC. European Commission, Brussels, Belgium.

European Commission (2002a). Regulation (EC) No. 178/2002 of the European Parliament and of the Council of 28 January 2002 laying down the general principles and requirements of food law, establishing the European Food Safety Authority and laying down procedures in matters of food safety. European Commission, Brussels, Belgium.

European Commission (2002b). Communication from the Commission – The operating framework for the European Regulatory Agencies. COM (2002) 718 final. European Commission, Brussels, Belgium.

European Commission (2003). Regulation (EC) No. 1829/2003 of the European Parliament and of the Council of 22 September 2003 on genetically modified food and feed. European Commission, Brussels, Belgium.

European Commission (2005). Draft interinstitutional agreement on the operating framework for the European regulatory agencies. COM (2005) 59 final. European Commission, Brussels, Belgium.

European Commission (2012). Roadmap on the follow-up to the common approach on EU decentralised agencies. European Commission, Brussels, Belgium.

European Commission (2013). Regulation (EU) No. 1271/2013 of 30 September 2013 on the framework financial regulation for the bodies referred to in Article 208 of Regulation (EU, Euratom) No. 966/2012 of the European Parliament and of the Council. European Commission, Brussels, Belgium.

European Commission (2015). Progress report on the implementation of the common approach on EU decentralised agencies. COM (2015) 179 final. European Commission, Brussels, Belgium.

European Food Safety Authority (EFSA) (2011b). Policy on independence and scientific decision making processes of the EFSA. EFSA, Parma, Italy. Available at: http://tinyurl.com/gshs4n3.

European Food Safety Authority (EFSA) (2014). Decision of the executive director of the EFSA on declarations of interest. EFSA, Parma, Italy. Available at: http://tinyurl.com/j3uvj7a.

European Parliament, Council and Commission (2012). Joint Statement and common approach of the European Parliament, the Council of the EU and the European Commission on decentralised agencies. European Union, Brussels, Belgium.

European Union (2015). Directive (EU) 2015/412 of the European Parliament and of the Council of 11 March 2015 amending Directive 2001/18/EC as regards the possibility for the Member States to restrict or prohibit the cultivation of genetically modified organisms (GMOs) in their territory Text with EEA relevance. Available at: http://tinyurl.com/y7fa2kfq.

Groenleer, M. (2009). The autonomy of European Union agencies. Eburon, Delft, the Netherlands.

Quirk, P.J. (1981). Industry influence in Federal Regulatory agencies. Princetown University Press, New York, NY, USA.

Salvador, M. (2002). Independent authorities. Un análisis comparado de los Estados Unidos, el Reino Unido, Alemania, Francia y España. Ariel, Barcelona, Spain.

Shapiro, M. (1996). Independent agencies: US and EU. Jean Monet Chair Papers 34. European University Institute, San Domenico, Italy.

Tarrant, A. and Kelemen, D. (2007). Building the Eurocracy: the politics of EU agencies and networks. Montreal, Canada.

Vírgala, E. (2011). Las agencias reguladoras de la UE. Comares, Granada, Spain.

Wonka, A. and Rittberger, B. (2009). How independent are EU agencies? RECON Online Working Paper. Available at: http://tinyurl.com/hwjfy3x.

Wonka, A. and Rittberger, B. (2010). Credibility, complexity and uncertainty: explaining the institutional independence of 29 EU agencies. West European Politics 4: 730-752.

12. Towards a new regulatory framework for GMOs in the EU: will it be third time lucky?

L. Escajedo San-Epifanio
Professor of Constitutional Law and Law & Ethics in the Biosciences, Department of Constitutional Law and History of Political Thought, University of the Basque Country, P.O. Box 644, 48080 Bilbao, Spain; leire.escajedo@ehu.eus

Abstract

This work analyses the key elements of the new regulatory framework designed by the European Commission to replace those which in the 1990s and 2000s regulated GMOs. Particular attention is paid to its rationale and reception by different EU institutions, stakeholders and some NGOs. Though guided by the same rationale and almost identical in literal terms, the legislative reforms promoted to shape this new regulatory framework, i.e. those included in Directive 2001/18/EC and Regulation 1829/2003 encountered, as we know, very different responses. The former was accepted and came into force via Directive 412/2015, and the latter, however, was rejected outright, especially in the European Parliament (619 votes to 58). In order to understand the situation that has been created, this work compares the third regulatory framework for GMOs in the EU with its predecessors – the legal frameworks of the 1990s and 2000s – and with the response it has met in the European Parliament, the Agricultural Council, the European Economic and Social Committee and the Committee of the Regions, as well as the views of stakeholders, some NGOs and literature from different disciplines. This will lead us onto issues such as the risk management model, voting rules, the application of the transparency and precautionary principles or, amongst others, the scope of Member States to make use of the opt-out prerogative granted to them by the new legislation. All aimed at answering, in the final reflection, the question as to whether this regulatory framework will prove to be third time lucky.

Keywords: new regulatory framework on GMOs, opt-out, differential traits, harmonisation, risk-management, precautionary principle, compelling grounds

12.1 Granting Member States some flexibility to decide upon certain particularities linked to GMOs at national, regional or local level: a Solomonic decision?

In the Old Testament, the Book of Kings, there is a scene in which two women appear before King Solomon, both claiming to be the mother of a newborn baby (1 Kings, 3: 16-28). Solomon, after some deliberation, declared that he did not know who was in fact the child's mother and said that, in all fairness, the best solution was to bring a sword with which to split the baby in two. The appearance of the sword prompted a series of reactions; among them the horror of the child's true mother, who said she would rather see him in the hands of another than dead. In the light of these reactions, Solomon passed his mythical judgement. Interestingly, in everyday language the expression *Solomonic decision* is often employed to describe a kind of just *division (almost in half) between the parties involved*, but the fact is that Solomon never really appeared to entertain the idea of dividing the child. His lesson

Leire Escajedo San-Epifanio (ed.) **Towards a new regulatory framework for GM crops in the European Union**
DOI 10.3920/978-90-8686-845-2_12, © Wageningen Academic Publishers 2017

of wisdom lay in the merit of bringing into play a new element (in this case, *a threatening sword*) which, though symbolically it seemed to take to the most dramatic limits the consequences of the absence of agreement, curiously paved the way to a just resolution.

I wonder whether perhaps, with the path it has taken in the construction of the third regulatory framework for GMOs in the EU, the real intention of the EU Commission has been none other than to bring into play an image (that of the sword) that might illustrate to Member States the consequences of continuing with almost twenty years of lack of agreement and of moratoria and provide a way out of the regulatory impasse (Salvi, 2016). '*The use of genetic engineering in plants and in food and feed is a subject which divides opinion in the Member States*' notes the Commission (EC, 2015a), and although there are many spheres in which Member States find it difficult to reach agreement (immigration, financial controls, public debt), the fact is that making decisions in this area is '*very much the exception to the usual functioning of the EU comitology procedure as a whole*'. According to the Commission, in 2012 the EU comitology procedure in Standing Committee was applied in 1946 cases in the whole of the EU and, of these, only 82 votes resulted in '*no opinion*' and nine of these procedures were food uses referred to the Appeal Committee[33]. However, in the case of GMOs for agri-food use, since 2003 Member States *(have) never obtained a qualified majority in favour of or against* a draft Commission decision authorising GMOs, whether for cultivation or for GM food and feed. In other words, that the result has always been a '*no opinion*' at all stages of the procedure, both in the Standing Committee and in the Appeal Committee. Consequently, the Commission has systematically been put in a situation where it has to take a decision on authorisations in the terms established by the legislation (EC, 2015a) and without the support of Member States in relevant committees. The Commission avoids mention of the fact that, although in some cases this was public knowledge, it has also been under pressure from certain Member States and the European Parliament, which sought that with regard to the precept indicating that when a 'no opinion' is issued the 'Commission *may adopt* the draft implementing act' (EP, 2015), that *may adopt* be interpreted in the sense that the Commission could at its discretion decide *whether or not to adopt* the draft and, if so, *with what content.*

After attempting this with an initial fairly harmonised regulatory framework (Directives 219/90/EEC and 220/90/EEC) and replacing this in the 2000s with a framework *fully harmonised at EU level* (with regard to Directive 2001/18/EC) the Commission appears to have opted on this occasion for a solution that works in the opposite direction: partially return to Member States their competences with regard to GM crops, food and feed. There are many spheres of EU functioning in which in ordinary politics disagreements arise, some of them very serious, and the Commission, entrusted by EU functioning as a whole to be *guardian of the Treaties* and to ensure understanding, had never before promoted reform in this direction (see in this monograph López Basaguren). For that reason the solution proposed has been described as *truly exceptional* (EESC, 2015), a solution understood to be justified on account of certain particularities linked to GMOs.

[33] The corresponding figures for 2013 and 2014 are respectively as follows: 1959/53/28 and 1908/46/21. Given by the Commission at the DOC COM (2015) 176 final, note 22.

In practical terms, the partial devolution of the competence – referred to by the Commission as *flexibilisation of some aspects of decision-making regarding GMOs* – has been reflected in some articles of the two main pieces of legislation in the legal framework for GMOs which was created in the 2000: *Regulation (EC) No 1829/2003 on genetically modified food and feed* (henceforth Regulation 1829/2003) and *Directive 2001/18/EC on the deliberate release into the environment of genetically modified organisms* (henceforth Directive 2001/18/EC). During their processing in the legislative bodies, however, each proposal for amendment met with a different fate. In the spring of 2015 Directive 412/2015 was passed, *amending Directive 2001/18/EC as regards the possibility for the Member States to restrict or prohibit in their territory* the cultivation of GMOs that have previously been authorised by the European Union. This law came into force and allowed the 19 Member States that so wished to make use of their opt-out facilities, as explained by Francisco Romero Caro in another chapter of this monograph. The amendment to Regulation 1829/2003, however, which would have allowed Member States to restrict *the use of approved GM food and feed in their territory* (EC, 2015b) was rejected outright by virtually all the parties involved: the European Parliament (EPRS, 2015), the Agriculture and Fisheries Council, the European Economic and Social Committee (EESC) (EESC, 2015) and the European Committee of the Regions (CoR) (CoR, 2015), as well as by the stakeholders of the *EU food and feed chain*[34] and some NGOs[35]. Particularly significant is the case of the European Parliament which, following the recommendation prepared by its Committee on the Environment, Public Health and Food Safety (ENVI-Committee), rejected the Commission proposal by 619 votes to 58, with 13 abstentions.

The acceptance of one of the reforms and the rejection of the second was, to a certain degree, especially noteworthy, because in a literal sense the proposal for amendment of Regulation 1829/2003 basically *mirrored* – as regards GM food and feed – the amendment included with regard to GM crops in Directive 2001/18/CE. This rejection, moreover, has put on the table an assortment of interesting reactions and reflections of those involved with regard to their appropriateness and viability, as well as their potential impact upon the functioning of the EU and of the single market. This situation begs the question as to whether, ultimately, this open reflection will affect the already approved Directive 412/2015. In this impasse in which we find ourselves, I wonder – rather innocently maybe – whether perhaps the proposal for partial devolution, exceptional in itself and atypical in being promoted by the Commission, has an element of the *Solomonic*. For, given its short life, perhaps the proposal, rather than deepen the division, sought more to show the parties involved how far this path could be travelled and at what price. What is currently beyond doubt is the generation of a new level of consciousness of the dissension in relation to GMOs in the EU.

It is difficult to know what direction the EU will take in this and other issues (Poli, 2015), because GMOs apart, it is obviously not enjoying one of the best moments of its history. What this chapter proposes is a review of *the rationale* of this third Regulatory framework for GMOs and its premises, its core assumptions, along with everything (emotions, opinions, reflections) that the (rejected) proposal for amendment of Regulation 1829/2003 has caused to be expressed. Firstly, I will offer an analysis of

[34] Among other: European Association of Farmers and Agri-Cooperatives Copa-Cogeca, European Association of BioIndustries, – EuropaBio, European Association of the Agrosupply Trade – COCERAL, and European Compound Feed Manufacturer's Federation FEFAC.

[35] Friends of the Earth, Greenpeace and European Organisation for Organic Food and Farming (IFOAM).

the *rationale* of the third regulatory framework (Section 2) as laid out in different official documents, as well as of the various responses to this rationale. Then, in greater detail, I will explore the core assumptions and the premises of this regulatory framework, and the normative elements in which they are reflected, to present the response given to the reform proposal not in a general sense, but *as it has been worded.* This will be undertaken in two separate sections. The first of these (Section) will address what is meant/implied by the fact that the third regulatory framework proposed maintains authorisation processes that are *fully harmonised at EU level.* This will lead to consideration of the role of the European Food Safety Authority (EFSA), and questions such as the application of ethical criteria and of the precautionary or transparency principles, and the tensions over voting rules and legitimacy. In the second sectionwe will consider the scope of the opt-out facilities available to Member States, the compelling grounds that may justify their use and the practical measures via which they may be manifested. Finally, and in the light of what has been analysed, the work will end with a reflection upon the open question in the title of this volume the final section: will this third EU regulatory framework in the EU for GMOs prove to be 'third time lucky'?

12.2 The rationale behind the construction of the third regulatory framework for GMOs in the EU and of its reception, from a legal-constitutional perspective

12.2.1 Towards a sustainable and healthy development of biotechnology

One hundred and seventy-seven States represented in what came to be known as the *Earth Summit* (UNCED, 1992) highlighted the importance for the Earth's survival of achieving an *essential balance* between economic, social and environmental interests. This required, as was reflected in the commitments adopted, a transformation of forms of production, consumption, life and relations, and in this desire for transformation there was also a place for modern biotechnology. Biotechnology by itself, reflected *Agenda 21*, cannot solve all the fundamental problems of environment and development, but nevertheless it

> promises to make a significant contribution in enabling the development of, for example, better health care, enhanced food security through sustainable agricultural practices, improved supplies of potable water, more efficient industrial development processes for transforming raw materials, support for sustainable methods of afforestation and reforestation, and detoxification of hazardous wastes. Biotechnology also offers new opportunities for global partnerships, especially between the countries rich in biological resources (which include genetic resources).

Hence, in the definition of its programme areas y activities, *Agenda 21* specifically included the need to '*establish(ing) enabling mechanisms for the development and environmentally sound application of biotechnology*' (Escajedo San-Epifanio, 2010). The OEDC, the FAO or the European Commission – among many other institutions – also declared in the 1990s that biotechnology was a '*crucial and indispensable tool in the management of Biodiversity vis-à-vis the challenges of food, health and environmental protection*' of the new millennium (Van der Meer, 1999), and that, along the lines of what is established chapter by chapter *Agenda 21* with regard to human activities with an environmental impact, it was necessary to guarantee that their development, in relation to both *new* and traditional techniques, respected both human and environmental health.

The drafting of the first regulatory frameworks for biotechnology coincided with a period, in the latter half of the 1990s, in which the pressure of public opinion and a certain *legislative euphoria* resulted in an accelerated drafting of laws with regard to advances in Genetics and Molecular Biology, especially attentive to their promises and their perils (Byk, 2008; Faulkner *et al.,* 2012). Beyond the promises and perils of biotechnology and other technologies, a certain euphoria of this nature was also apparent in Environmental Law, and in particular in the wake of the agreements reached in the UNCED. The history of Science is full of discoveries that in their time were initially regarded as dangerous or unacceptable, but from the perspective of the Law never before had the perspectives or potentialities (rather than the tangible results) of a technical-scientific sector led to such a degree of anticipation by legal responses to social problems that had yet to materialise. In some way, an instrument that traditionally formalises solutions to social problems – Law – had presented itself as a mediator, an instrument of balance, a reference in order to find a harmonious balance that, looking to the future, might guarantee with regard to scientific advances a middle path between absolute fundamentalism and unprincipled arbitrariness (Eser, 2002).

This mediation role, more symbolic than tangible, has been frequently repeated in legal texts – it would be impossible to quote them all. Some legislative texts succeeded in this task, but others tended more towards fundamentalism or arbitrariness than towards the middle way. Given the number of countries involved in legal instruments of this kind, I shall highlight two instruments which, as well as attracting broad consensus, were I believe successful in expressing the challenge of finding a reasonable balance: on the one handed, those drafted in the UNCED, especially in the aforementioned *Agenda 21* (177 countries involved), which in Chapter 16 expresses the objective of *enabling mechanisms for the development and environmentally sound application of biotechnology*; and, on the other, the *Universal Declaration on the Human Genome and Human Rights*[36] (UDHG) (UNESCO, 1997), approved with the participation of 186 States and the official support of the USA (which is not a member of the UNESCO). In the UDHG, the commitment to balance is expressed in these terms: *'research on the human genome and the resulting applications open up vast prospects for progress in improving the health of individuals and of humankind as a whole'*, and should *permit* that research and practice the goal of which is the improvement (of the health of individuals and of humankind as a whole), *provided it is* fully respectful of dignity and fundamental rights. Verification of that *provided,* necessarily leads us to a case-by-case study of each advance in this area of Science and Technology.

In a region like Europe, the constitutional texts of the Member States and the actual *Charter of Fundamental Rights of the European Union* (CFREU) reflect how we have deposited in the freedom of science (art. 13 CFREU) and the freedom to conduct business in accordance with Community law and national laws (art. 16 CFREU) much of the promotion of scientific and technological development, and have also established public instruments through which society as whole supports and encourages this development. Consequently, unlike in the case of developing countries, the most

[36] Universal Declaration on the Human Genome and Human Rights, adopted unanimously and by acclamation at UNESCO's 29th General Conference on 11 November 1997.

visible aspect of the regulatory frameworks created in relation to the applications of Genetics and Molecular Biologyhave been the administrative mechanisms that seek to *promote use and prevent misuse or undesired adverse effects* that these advances may have upon dignity or fundamental rights, as well as upon the environment. Thus, and with regard to GMOs, starting with the first regulatory framework created for these in the EU in the 1990s, a system was established which decides *case by case* upon authorisation requests for restricted use or deliberate release of a GMO in the EU (in Directives 90/219 and 90/220), but assumed as a rationale of the law that biotechnology is a sector from which the EU expects useful contributions. Along with these general laws, others of a more sectorial nature were developed vis-à-vis GMOs, such as those relating to their use in the production of medicine, feed and food or products for agricultural use, such as seeds. This is the case of Regulation EC No. 2309/93 on *medicinal products for human and veterinary use* or Regulation EC No. 258/97 *concerning novel foods and novel food ingredients*, and Directive 98/44/EC on the legal protection of biotechnological inventions. We can also find some references to GMOs in the law on inland transport of dangerous goods and the one regulating the exposure to biological agents at work.

12.2.2 The core assumptions of the EU frameworks for GMOs

Amongst the considerable number of ethical and political issues arising from GMOs referred to by other authors in this monograph (Kaiser; Bruce; Levidow), the decision of Member States to agree upon, together, a framework for this area in the institutional sphere of the EU meant the tacit acceptance of a series of core assumptions or, in other words, a specific stance with regard to some of the ethical and political issues linked to GMOs. This is a practically mechanical consequence of the fact that Member States decide to project, also on products that are or that contain GMOs, the cooperation agreements they have assumed, in general, via the Treaties and successive reforms that have come to form what is today's European Union. Let us examine this in greater detail.

Including biotechnological policies in community policies implies, as an initial core assumption, locating them in the context of a supranational entity (the EU) which emerged, fundamentally, for market integration. The EU, it is often said, is *more than a single market,* and that is true; but, even if it is more, the EU is, *also*, a free market area. In this space, and as occurs internally in each and every Member State, something as essential for human survival as food is regarded as a mere commodity, as too are other key elements that along with food form part of the minimum subsistence enshrined in Article 25 of the Universal Declaration on Human Rights (for example, housing). The guarantee of the human right to food and food security, the right to preserve the autonomy of food systems are values recognised to a certain degree as part of European legal culture, but in the wording of EU law and the laws of the Member States there are no legal measures solid enough to put forward a defence of these values that *opposes or limits market freedom*. It is said that globalisation has substantially modified our food systems and the manner in which decisions are made regard to the production, distribution and consumption of food, although to a large degree what it has done is further develop a model that the EU had already initiated. Free trade, basically, results in the establishment of high quality and safety standards for products, particularly high in the case of novel *food* and GM food and feed, but there is barely room to attend to other considerations. Justice and equity seem to most of us to be very important values, but in legislative terms at least, for the time being we Europeans appear to have agreed to live in a regulatory framework in which little in the way of justice or equity can be demanded of the market. It is difficult, then, to expect, with regard to the GMO market in particular, values to prevail that, so far, we have relegated in the regulatory framework of our single market.

As has been discussed elsewhere in this work, the EU model of food governance and, broadly speaking that of Member States too, is not receptive to proposals for transformation of the agri-food system that seek to adapt it to parameters such as equity or justice (Escajedo San-Epifanio, 2012, 2015). Thus, and though it is perfectly legitimate to hope that the regulatory framework for GMOs might represent, only for these products, an exception to this market philosophy and to this way of conceiving of the status of foods, defence of this stance must bear in mind that its materialisation will, in any case, come up against the laws that within the Member States, in the EU and at international level have been created to protect market freedom.

Another factor to be taken into account is a factual rather than legal restriction. Although there exists on paper a legislative embodiment of the commitment to a single market, its effective implantation has neither been nor is proving to be trouble-free within the EU, and this is perhaps even more evident in the agri-food sector. Since the *Spaak Report* (1956), one of the starting points of what would be the European Economic Community, it was considered to be essential that economic integration *also embrace* this sector (McMahon, 2007), but it has never been easy. Neither has it been easy at an international level. In the Uruguay Round (1986-1995), the member countries of the World Trade Organization (WTO) drafted an *Agreement on Agriculture* (AoA) which, among other things, included a 'peace clause' (art.13 AoA) or commitment to reduce agricultural disputes to a minimum. Because, while they all sought to obtain benefits from free international trade in agri-food products, they also employed economic and non-economic obstacles to protect themselves from the possible negative consequences of this trade. The peace clause has not brought, far from it, peace to this sector and, whilst GMOs have not been the primary cause of these disputes, the dramatic appearance on the scene of some GM crops (which, for instance, in the case of soya and cotton substantially exceed the percentage of conventional crops) has clearly placed them at the heart of much of the tension.

Finalising or reforming any agreement that affects agri-food trade is, by its very nature, something extremely complex, and, in the dialectical confrontation between the opposing parties, neither statements nor silences can be interpreted independently of the tension inherent to the situation. In my opinion, there appears to be an illusion to this consideration in the EESC position paper on the proposal for reform of the third regulatory framework for GMOs (EESC, 2015). If we look closely, before offering a detailed opinion regarding the specific proposal, the EESC warns that the Common Agricultural Policy (CAP, which encompasses agriculture, livestock and fisheries) is in a critical situation. Due to the current CAP's operational and budgetary constraints, states the EESC, '*many countries are now questioning the competences conferred by the Treaties in this area, causing political damage to the European institutions*' (EESC, 2015). For this reason, in its position on the new regulatory framework it expressly declines to participate in a debate over whether GMOs are inherently 'good' or 'bad', and reflects only upon the question of whether the framework proposed is of sufficient quality to serve, in the field of GMOs, the commercial cooperation project in which Member States are immersed. In this intrinsic or inherent consideration of GMOs there is no consensus, hence its cautious stance, and it also warns that when it comes to undertaking reforms that affect the EU agri-food sector neither the legislators nor the EESC should ignore the fact that some Member States that are not '*natural*' beneficiaries of the CAP are strongly opposed to the latter.

Theoretically there is always the possibility of, on the one hand, reform of the Treaties with such drastic decisions as excluding from them the CAP or, on the other, as finally seems to have occurred with Great Britain, specific Member States abandoning the framework of the Union, but while – we hope – these paths are not taken, the regulatory proposals that affect the agri-food sector must move and seek common ground amidst the unease generated by this swell.

Until such time as they are reformed, the Commission's task of formulating a new regulatory framework for GMOs must, necessarily, respect the parameters included within the Treaties. In this case and in the short term, it will be possible to channel many of the proposals in reasonable fashion via labelling systems (be they voluntary or obligatory) that allow consumers to exercise, by way of *vote*, their shopping-bag power. Though in relative terms and in comparison with what some aspire to, this may appear little, the fact is that it far exceeds what occurs in most regions of the world, where products that are or that contain authorised GMOs carry nothing identifying them as such. In the light of its general food law, it is believed, for example in the US, that with regard to these products there is no specific risk or any other type of *differential element* or particularity which justifies requiring that producers and distributors assume the costs involved in labelling. In the EU, however, this information is considered to be relevant in order that consumers may exercise their options and, in fact, an additional requirement has-been included that, if we compare it with labelling requirements for non-GM food and feed, represents an important exception within EU *General Food Law* as a whole. I refer to the fact that with regard to agri-food products containing GMOs, apart from the wording 'this product contains GMOs', the law requires that consumers be advised as to whether the GM food in question '*may give rise to ethical or religious concerns*' (art. 13 R. 1829/2003). As this provision has yet to be put into practice academic literature and jurisprudence have yet to pay it much attention, but it is difficult to argue that this requirement is based upon any *differential trait* or particularity compared with non-GM products, because requirements of this nature have not been included with regard to other non-agri-food products, such as medicines or textiles[37], which may contain GMOs and are also commercialised in the EU. I am more inclined to think that it is a requirement born of the fact that the Commission, as I have said, describes in these terms: '*the use of genetic engineering in plants and in food and feed is a subject which divides opinion in the Member States*' (EC, 2015b).

To summarise, and in the light of the above, the legislation regulating the restricted use and deliberate release GMOs generically accepts the use of the latter in different sectors of economic activity and offers them a similar status to that of any other scientific or technological advance, opening the way for the resulting products to enter the market provided they comply with a series of strict safety and quality standards. Supervision of which GMOs and resulting products comply with the standards demanded is a question that, given its highly technical nature, is referred to a Standing Committee (in which Member States participate) but assisted by advisory bodies formed by independent experts. Specifically, and on a mandatory basis, with regard to GM crops, food and feed the intervention is required, at the very least, of the EFSA, although there is also room for the participation of competent national authorities and other experts. In addition, and in relation to other non-scientific

[37] For example bearing in mind that 75% of the cotton grown in the world's transgenic according to ISAAA Brief 51-2015.

considerations that the Standing Committee wishes to take into account, there is also provision for possible consultation with the *European Group on Ethics in Science and New Technologies* or any other appropriate body it might establish, with a view to obtaining its opinion on ethical issues (art. 33, R. 1829/2003).

The regulatory framework of the 1990s, the one implemented in the 2000s (via Directive 2001/18/EC and Regulations 1829/2003 and 1830/2003) and the new regulatory framework, currently in an impasse, reflect this line. There is some disagreement with this manner of articulating the relationship between science, society and governance with regard to GMOs (see Levidow'sdetailed contribution to this volume), but from the legal-institutional point of view this decision-making framework is, essentially, coherent with that which the EU applies to every type of scientific or technological advance. In the few aspects in which the regulatory framework differs from legislation on non-GM products, especially in the question of labelling, the difference seeks to facilitate decision-making for those who wish to make use of their freedom of thought, conscience or religion (art. 10 CFREU). In comparative terms, and notwithstanding other considerations that Professor Arroyo Aparicio offers in this work in relation to labelling, there are no other food products in the EU – or even non-food GM products – with respect to which the consumer is afforded so extensive a right to information (Raybould and Mitchell, 2003).

12.2.3 The essence (or synthesis) of the European Commission's proposal for the third regulatory framework

The dissatisfaction with the regulatory framework for GMOs was considerable on the part of both those who defended the use of GMOs in the agri-food sector – especially from the scientific-technological sector and amongst stakeholders – and those opposed to such use and, consequently, a welcome was given to the Commission's decision to drive a reform that many regarded as necessary. It must be said, therefore, that the widespread rejection of the proposal, particularly the amendment of Regulation 1829/2003, was not because it was considered inopportune, but due to the specific terms of the preparation and drafting of the reform.

In essence and as I have indicated, the Commission's proposal maintained as *fully harmonised the authorisation procedure of GMOs and GM food and feed at EU level*, and in coherence, retains most of the literal content of the two key provisions of the second regulatory framework (Directive 2001/18/EC and Regulation 1829/2003). It only modifies certain specific points. This is the initial basis or premise of the proposal, which I will analyse in detail in the next section. This basis having been established, the task of legislative revision focuses on introducing into both provisions a *flexibilisation mechanism* enabling Member States to address some of the concerns posed by GMOs and GM food and feed. This is achieved, in practice, by the inclusion in both provisions of an opt-out prerogative for Member States. The opt-out, however, does not represent a break with the first of the premises, in other words, with the maintenance of the authorisation procedure that is fully harmonised at EU level. This is why the proposal is said to introduce *some flexibilisation* regarding '*certain particularities linked to GMOs*'.

The means of legally articulating this flexibilisation has been that of guaranteeing an opt-out prerogative, an area of free will, within a series of requirements that guarantee: firstly, the integrity of the single market; secondly and linked with this, respect for that full harmonisation of the authorisation procedure (including risk management); and, thirdly, a series of guarantees of respect for article 41 CFREU, which recognises the *Right to good administration*. As things stand, as we have seen, this legislative model has been accepted with regard to Directive 2001/18/EC in relation to the use of GM crops, but not with regard to Regulation 1829/2003. Hence the complex situation, at present, vis-à-vis the culmination of the third regulatory framework for GMOs in EU. Let us consider the details of the proposal and the elements that have aroused opposition.

12.3 The premise of preserving as fully harmonised the authorisation procedure for GMOs at EU level: precautionary principle, and ethical and legitimacy concerns

12.3.1 A fully harmonised authorisation procedure at EU level involves maintaining a single risk-management system, based on independent risk assessment

Amongst the core assumptions of the regulatory framework that it proposes for GMOs, it is important to underline that the Commission considers that it is important to maintain a single risk-management system, based on independent risk assessment in preference to a system involving national authorisations with mutual recognition. It alludes thus to one of the most significant reforms incorporated in the 2000s vis-à-vis the regulatory framework for GMOs established in the 1990s. That reform, it is worth remembering, was not the consequence of anything connected with the safety of GMOs, but as the result of a series of food safety crises related to other products (in particular, the case of BSE or *mad cow* disease) which threatened to undermine consumer confidence in the single market for food. A single risk-management system, said the Commission, is the most effective way of ensuring the same level of protection throughout the EU, as well as the functioning of the internal market, and not only with regard to GMOs but in relation to all agri-food products. It considers, logically, that no reasons exist for decision-making on GMOs to be rendered an exception, with risk that this might produce distortion or deterioration of the high level of protection currently established.

This configuration reinforces and specifically supports the fundamental role in the risk-management system of the EU agri-food chain played by the EFSA – see in this work the chapter by Salvador Martínez – an institution which, along with the Commission, has been the main target of all the reactions, emotions and criticism that, in one sense or another, have resulted in several decades of disagreements over GMOs in the EU.

12.3.2 Scientific risk assessment, risk management and precautionary principle

In the decision-making process on the authorisation of GMOs and GM food and feed in the EU there is one phase of scientific risk assessment and another of management of the latter. The technical-scientific evaluation, which revolves around the EFSA but allows the participation of Member States and other interested parties, is based upon a list of technical requirements and some annexes detailing these, requirements established in their entirety by legislative bodies of the EU (i.e. Parliament, Commission and Council). This means that the function of the EFSA with regard to decision-making on GMOs is clearly restricted: it must check that a request for authorisation of GMOs or use of the latter satisfies the strict requirements established by the legislative bodies. In fact, if something stands out with regard to EU legislation on GMOs in comparative Law it is the rigour demanded in relation to the control of risks to human health and the environment and the detail with which this is specified (Escajedo San-Epifanio, 2008, 2010; Lucht, 2015; Varzakas, 2007). In this sense, and from the legislative point of view, it is easy to respond to criticism like that voiced by those who reproach the EFSA for not taking into account 'other social considerations' (a criticism which, amongst others, is expressed by the CoR). Addressing these considerations, simply, would have meant overstepping the mark in the exercise of its functions. Its task, after all, is not a mechanical one but rather, as occurs in all those cases in which in the EU (and in the Member States) it has been necessary to create a consultative body in the technical-scientific field, it must offer a response to what is asked of it employing, to this end, the scientific knowledge available and as many verifications as are considered necessary. This, necessarily, involves providing the assessment body with a margin of organisation of the way in which it will fulfil its function with a margin of appreciation consistent with its expertise, but without departing from, in this case, the sphere of scientific-technical assessment.

Upon completion of this assessment, risk management is assigned primarily to Member States (meeting within the Standing Committee or the Appeal Committee), a task the exercise of which is facilitated via a *decision draft* prepared by the Commission on the strength of the EFSA report. The Member States, meeting in the above mentioned committees, may deviate from the decision draft and take into account other criteria than those assessed by the EFSA, but the fundamental right to good administration (art. 41 CFREU), and its projections in both the Treaties and in the legislation applied to GMOs will require of the Standing Committee that it explain its reasons with clarity. Thus, and in the light of art. 41 CFREU and other legal provisions, the European Court of Justice may, if necessary, check whether those criteria not included in the law by the legislative bodies but which the Standing Committee has seen fit to apply, have been applied in a manner compatible with the principle of proportionality and non-discrimination. Only as a last resort, in the case of no opinion reiterated in both committees, does the law state that the decision will rest with the Commission. Given its institutional nature, the EFSA, unlike the Committees in which the Member States participate, is assigned by the legislation on GMOs no amount of discretion to address, among others, possible political, social and/or ethical issues or to apply a precautionary approach.

Given this distribution of responsibilities in the risk assessment and management process, there does not appear to be much point seeking to delegitimise an advisory body – in this case the EFSA – for not entering into the evaluation of aspects (e.g. ethical or social) that are not within its area of competences or the Commission for not exercising an amount of discretion to which it is not entitled. If anything, the question that should be asked is why Member States, instead of working on the development of these criteria, have preferred repeatedly to resort to a *no opinion* in the Standing Committee and the Appeal Committee. This is a very convenient stance, it cannot be denied, for Member States, because although the precautionary principle appears to invoke something obvious (apply caution), adopting measures based upon it is more difficult than might appear. This is explained by two of the most important institutional documents issued to date with regard to this principle: one by the Commission (EC, 2000) and another by the UNESCO (2005). The precaution, explains the Commission, is an 'eminently political' decision and its criteria are *far from clear* (2005). In fact, the application of a precautionary approach must deal with the circumstance that, in the framework of democratic systems, any actions, even precautionary ones, have legal constraints (O'Riordan *et al.*, 2001). In the case of the EU, it is the treaties and the *legal* provisions, along with the community acquis that delimit that legal constraint. In the case of *T-13/99 Pfizer,* the ECJ recognised that *political decisions such as precautionary ones are not very amenable to judicial review*, but this does not mean that it is not possible to scrutinise them from a legal perspective. Because it is essential to prevent arbitrary decision-making *disguised under precautionary clothes* (Alemanno, 2013), which would violate the fundamental right to good administration (art. 41 CFREU).

To be consistent, when the Community administration proposes the adoption of a specific measure based on the precautionary principle (for example, the non-authorisation of a GM crop for some reason different from non-compliance with the requirements established in the law), such a decision must be justified. If the reason were the application of a precautionary approach, the accompanying justification must explain the manner of the application to the case of the general principles of risk management (EC, 2000): (1) proportionality between the measures taken and the chosen level of protection; (2) non-discrimination in application of the measures; (3) consistency of the measures with similar measures already taken in similar situations or using similar approaches; (4) examination of the benefits and costs of action or lack of action; (5) review of the measures in the light of scientific developments. This decision formula is open to judicial review.

12.3.3 Transparency, accountability, legitimacy and voting rules in the process of management of risks associated with GMOs

Transparency, accountability or democratic consent are, along with others, important values of public policy that repeatedly arise in public debates on issues that, though diverse, have in common a high level of conflict. An analysis by the Berlin-Brandenburg Academy of Science (Bratsch, 2014; Müller-Röber *et al.,* 2013) explained that, amongst the peculiarities of the way in which the discussion of GMOs is approached, as in debating agricultural biotechnology, each side also tends to accuse the other of anti-morality and there is also frequent use of expression like the lack of legitimacy of the decision-making bodies, the consultative ones included. Thus, for example, the European Parliament Research Service report (2015) reflects that '*in recent years there has been criticism from NGOs, the media and the European Parliament about the Independence of the EFSA risk-assessment procedure*' and that '*it has been suggested that there are conflicts of interest with members of EFSA's scientific panels having overly close ties to industry*'. Also referred to, as evidence of a lack of transparency, is the fact that critical information contained in the research sea often classified as 'confidential'.

Respecting freedom of opinion, if we consider the current regulatory framework for GMOs in the EU in the light of common principles in European legal-institutional culture, the terms transparency, democracy or legitimacy (and the absence of these) are not always employed in an appropriate manner in some of the suggestions referred to. Thus, to begin with, we find that the main EU institutions (Parliament, Council and Commission) and the Member States, each of these with the legitimacy that corresponds to it, set out in the Treaties a system of legislative production and, subsequently, the citizens of the Member States (via referenda or decisions by national parliaments) endorsed the EU institutional model with a high level of consensus. Irrespective of whether or not one agrees with the decisions taken in the day-to-day functioning of the EU, from the legal-institutional point of view the legitimacy of this functioning cannot be questioned.

In this decision-making context, and with full legal guarantees, the law regulating GMOs in the EU was passed with the same legitimacy as all the other legal provisions in force at EU level. Directive 2001/18/EC and Regulation 1829/2003 contained, in turn, a decision-making process regarding authorisation requests in which the legislative bodies have granted a series of functions to the EFSA and others to the Commission, Standing Committee and the Appeal Committee that are not exceptional in comparison with authorisation models that at EU level are used to decide upon other authorisations. Moreover, and in any case, actions performed in application of this law are subject to the corresponding jurisdictional control. Obviously, as occurs with the specific decisions adopted in the Parliament or Government of a Member State on many issues, decisions resulting from decision making processes on GMOs may be more or less popular with those affected by them, but, once again, I must insist that from a legal-constitutional point of view it is not correct to speak of a lack of legitimacy or democracy unless there is a violation of the rules of the decision-making process or there is the case of some element – e.g. unresolved conflict of interest or arbitrariness – generating dysfunctions in a specific decision-making process.

The *suggestion* of conflicts of interest in relation to bodies asked to issue independent rulings is an extremely grave accusation and should be taken very seriously. Very seriously, on the one hand, to check, obviously, if the suggestion is true and therefore it is necessary to take action. But no less important, moreover, is the fact that these kinds of suggestions must be analysed in order to check, also, if they are being employed to generate a threat to the external independence of the advisory body. In other words, to check whether an independent consultative body is being put under improper pressure by stakeholders, pressure groups, the media or even the executive or legislature, including the Member States. If this were so, it would be necessary to implement measures to protect its correct functioning, in the same way that judicial independence is reinforced when cases are tried under intense scrutiny by pressure groups (regardless of their opinions or ideology). Advisory bodies, all of them, need internal and external protection from outside influence or control, even possible influence exerted by the body or institution which they advise.

On a separate issue, some Member States had proposed changes to the weighting of votes in the two committees referred to above (Standing Committee and Appeal Committee), rather than having a national prohibition system, as is noted by the CoR (par. 44). Depending on how the vote were modified, of course, this would have meant allowing some Member States to extend their desire for national prohibition to the whole of the EU and thus rendered unnecessary monitoring on the borders of trade restrictions. In line with the opinion maintained by the CoR (CoR, 2015), this institution finds this to be the most *practical* option. But what is not taken into account is that to operate in this way would have, however, a significant impact upon the functioning of the EU. The Treaties, for understandable reasons, cannot allow each Member State to promote a modification of voting rules to suit its own interests at any given moment, and the use of new techniques in the agri-food sector should not be an exception.

With regard to transparency and, coupled with it, risk communication, the model prepared by the EU vis-à-vis GMOs is that of the publication of all information that is not considered 'confidential'. This latter possibility, of confidentiality, is exclusively reserved to those cases in which it is regarded as appropriate to protect certain information and some data in relation to applications in order to protect the investment made by innovators. What is sought here is a balance between two values that socially and with regard to all technical-scientific advances have been considered relevant in the EU: on the one hand, transparency; on the other, support for those who promote research that contributes to socio-economic development. The absence of publication of all the information, nevertheless, does not mean that there are elopements of a product or technology that are exempt from risk analysis by the body that performs scientific-technical assessment, or from the risk management that is undertaken by the administrative bodies. These bodies do have access to all confidential information.

Meanwhile, and also in connection with risk communication, on some occasions the latter is employed as a synonym for *direct public participation* in decision-making with regard to risks; in other words, it is suggested that risk communication should take the form of the establishment of some kind of public platform that, in binding manner, participates in the final phase of decision-making. But risk communication and public participation in decision-making with regard to risks are two different concepts, both with respect to GMOs and in relation to processes of managing other environmental risks. The regulatory framework in force in the EU establishes, certainly, a system of advertising and mechanisms enabling citizens individually or collectively to send information and opinions to the risk managers, but there is no provision for activation of consultation platforms of a more or less binding nature. Though this is not the occasion to elaborate on this, in academic literature one finds both views in favour of creating these kinds of platforms and voices warning that, in cases arousing a high level of political controversy, the creation of such platforms may distort both the criteria regulated as a basis for decision-making and the way in which society is really represented in this process (Cook and Lewandoski, 2011; Sunstein, 2005).

12.4 The opt-out prerogative assigned to member states in the third regulatory framework for GMOs in the EU: its nature, content and limits

From the point of view of Constitutional Law, the proposal for flexibilisation contained in the *Communication from the Commission Reviewing the decision-making process on genetically modified organisms* (EC, 2015a) is an interesting exercise in legal construction in the sphere of the European Community, in that the flexibilisation in question seeks to *devolve to Member States* certain competences that within agri-food law as a whole – and not only therefore in relation to GMOs – are EU competences. Partial in nature, that devolution, moreover, only occurs with regard to certain peculiarities linked to the use of GMOs. Once the idea has been clarified, putting it into practice involves a genuine exercise of creativity.

I shall concentrate on three points. Firstly, and given the turn of the screw represented after twenty years of moratoria by this devolution of competences, I shall take heed of the voices that, comparing our regulatory framework with that of other legions of the world, advocate a return to the 'differential aspect' that led in the 1990s to the beginning of an exceptional trajectory for GMOs in agrobiotechnology and the consequences of that decision. Secondly, I shall explore the definition of the peculiarities allowing for exercise of the opt-*out* option, that is, the compelling grounds admitted by the third regulatory framework. And finally, I shall analyse the specific *measures* –restrictions 'on use' – which in accordance with the legislation may be adopted upon the basis of these grounds.

12.4.1 Revisiting the controversy over the differential 'aspects' or particularities of GMOs and the way those aspects should be taken into account

When the regulatory framework in force in the EU since the 1990s is compared with those applied in other regions of the world, especially in the country that is today's main producer at a world level (the US), emphasis is placed upon the difference between *process-based* and *product-based* decision-making (Levy and Newell, 2000). But equally more influential than that difference have been other important decisions, amongst them the decision as to what should be the headquarters of agrobiotechnological policies in the EU. The headquarters of biotechnological policies in the EU was initially the *Directorate-General for Science, Research and Development* and following various debates between the directorates concerned, with the formulation of the 1990s Directives on GMOs eventually assigned leadership of biotechnological policies to the *Directorate-General for Environment* (Von Homeyer, 2002). This DG is, comparatively, less susceptible to the influence of stakeholders in the food and feed chain than in other regions of the world and at the same time, is subject to greater influence from environmental groups or other forces (Bernauer, 2003; Patterson, 2000; Rosendal, 2005; Von Homeyer, 2002). Along with these two important factors, in my opinion there are others that have also had a significant influence upon the zigzagging evolution of biotechnology in the EU, such as, on the one hand, the lack of cohesion in the sector – due to the diversity in the size and the profile of the companies which form it – or the significant collateral impact of the tensions that, in relation to *non-GM* products, have arisen in both the CAP and in the agri-food market as a whole.

In the light then, of the circumstances in which the EU finally decided to create a specific (and different) regulatory framework for agrobiotechnological products, led moreover by an atypical Directorate given the trajectory of the sector thus far, it is understandable that, as it stated by the heading of this section, twenty years later there is a return to both the *differential traits* or particularities upon which the construction of an exceptional system for the GMOs is based, and to that decision regarding competences. Some believe it is time to review the entire regulatory framework 'to allow international harmonisation and to avoid trade disruptions between countries and continents should take the new trait/product into account and not the technology to generate it' (Shiemann and Hartung, 2015). Pressure in favour of this international harmonisation has been increasing during the last decade, on the one hand via the challenges filed by some countries against the EU (Section 12.5.3) and on the other through recent consensus documents being prepared by way of a summary of twenty years' experience of commercial GM crops. Amongst these, of particular interest are two documents by the Task Force on Safety Assessment of Foods and Feeds derived from Transgenic Crops de la OECD[38]. These concluded that the scientific evidence to date is that GM varieties are *compositionally equivalent to conventional varieties*, and therefore there is no difference between GM crops and conventional crops.

There is also a return to the question of what it is it that is different about GMOs which justifies a different treatment compared with that of *non-GM* products and the consequences various ways of this differential policy, as for instance, from the perspective of scientific and technological

[38] Safety Assessment of Foods and Feeds Derived from Transgenic Crops, Volume 1 and Volume 3, April 2015. Online Access at the OECD library.

development (see among many others: Ammann, 2014; De Francesco, 2013; Klümper and Qaim, 2014). The *plurality views* that exist (see, in this work, the chapters by Kaiser, Bruce, Reddy) and the tension between these appear still to represent an obstacle to achieving consensus. In this scenario, in any case, it does seem to me worthwhile highlighting the fact that *Agenda 21* of the United Nations, in Chapter 16, neither draws a dividing line between modern and traditional Biotechnologies, nor limits to the former the goal of being *environmentally sound.* 'Biotechnology is the *integration* of the new techniques emerging from modern biotechnology with the well-established approaches of traditional biotechnology', declares the Agenda, and there is repeated reference to the use of the combined resources of both, establishing the objective of environmentally sound management in general. In other discourses, however, the narrative assumes that there exists a clear and firm dividing line between the traditional and the new (Pollock and Hails, 2014).

To this debate has been added the question as to the status that corresponds to New Plant Breeding Techniques (NPBTs) (Spring and Eriksson *et al.,* 2016). Developed over the last decade, these make it possible to modify plant genomes in a more precise manner than previous techniques and if they are notable for something with regard to transgenesis, it is for the fact that they do not introduce into organisms *foreign DNA.* In order to analyse whether or not these techniques have been addressed within the turbulent regulatory framework for GMOs we need to check where EU legislation situates the line between new and traditional techniques, because though the term GMOs is used more frequently in everyday language, Directive 2001/18/EC employs the expression to refer *only to some GMOs.* Thus, as far as the Directive is concerned 'genetically modified organism' means '*an organism, with the exception of human beings, in which the genetic material has been altered in a way that does not occur naturally by mating and/or natural recombination*' (art. 2), but, and this is very important, the terms of this definition are completed via technical annexes (IA, with two parts, and IB).

In order to define the scope of application of the GMO regulatory framework in the EU, it is stated in Directive 2001/18/EC that '*genetic modification occurs at least through the use of the techniques listed in Annex I A, part 1*' and that, by contrast, '*the techniques listed in Annex I A, part 2, are not considered to result in genetic modification*', despite the fact that, strictly speaking, they are techniques that do genetically modify the organisms to which they are applied. If we compare the two lists in Annex I A, we find that the techniques included are those that involve *the insertion of nucleic acid molecules* produced by whatever means *outside an organism*, those that involve *the direct introduction into an organism of heritable material prepared outside the organism (e.g. by micro-injection)* or cell fusion or hybridisation techniques in which *new combinations of heritable genetic material are formed by means that do not occur naturally.* However, other processes, such as conjugation, transduction, transformation or polyploidy induction in which recombinant nucleic acid molecules are not used, are excluded from the regulatory framework specifically for GMOs, being applicable to the general framework for agri-food products.

This does not conclude, however, the technical definition of the GMOs which remain within the scope of application of the so-called *EU regulatory framework for GMOs*, because what Annex IB does is specifically *exclude* a series of techniques/*methods of genetic modification*: (1) mutagenesis; and (2) cell fusion (including protoplast fusion) of plant cells of organisms which can exchange genetic material through traditional breeding methods. Of the techniques excluded, perhaps the most interesting is the case of mutagenesis, as this involves genetic modification of the organism employing chemical compounds or irradiation application; this technique does not quite appear to fit the definition of *naturalness* which some voices assign to organisms that are *not GMOs*. In fact, Regulation EC 834/2007 on organic production and labelling of organic products in the EU excludes the possibility of labelling as ecological both GMOs in all their forms, and products treated by ionising radiation.

NPBTs, as explained by authors Casacuberta, Nogué and Du Jardin in another chapter of this monograph, are in a kind of limbo, in spite of the fact that since 2007 both the European National Competent Authorities and the Directorate-General for the Environment of the Commission have organised working groups on different aspects related to these techniques – state of science for adoption, economic impact, possibility of detection, etc. (Shieman and Hartung, 2015). The two contrasting visions of NPBTs are clearly synthesised in two documents from 2015, on the one hand the *European Academies Science Advisory Council(EASAC) Statement* and, on the other, an *Open letter to the Commission* (NGO coalition, 2015) signed by a coalition of eight NGOs[39]. The EASAC requests that the EU policy development for agricultural innovation should be transparent, proportionate and fully informed by the advancing scientific evidence and experience worldwide, suggesting a review, in the light of new knowledge and experiences, of some political decisions adopted in the 1990s. Specifically, EASAC demands to resolve current legislative uncertainties and asks EU regulators to confirm that the products of NPBTs, when they do not contain foreign DNA, do not fall within the scope of GMO legislation. By contrast, the Open Letter of the coalition of the eight NGOs referred to states that the '*same principles* 'currently applicable to GMOs '*must apply to all new genetic engineering techniques and applications*'.

Still pending too, on another note, is the different treatment received by those GMOs used in the agri-food sector and those employed in other sectors (such as pharmaceuticals or textiles). *De lege*, as one would expect from a *process-based approach*, applicable to all of them is Directive 2001/18/CE and the criteria established herein, but the specific Regulations, for example that for medicine for human and veterinary use, have reformed the Directive excluding these products from *the decision-making process provided for in Directive 2001/18/EC* and replaced it with an alternative model, which has generated less controversy. In my opinion, it will be difficult to argue that the 'differential aspects' or particularities which justify a different treatment for GMOs are due to scientific-technical questions (including risks) when GMOs generated in the same way are treated differently.

A priori, the significant differences between Member States in the field of Biomedicine were regarded as potentially more of a threat to the maintenance of market unity than the applications of genetic engineering in sectors like agriculture or food. Whilst in 1991-1992 74% of EU citizens *tended to agree or totally agreed that genetic engineering research on plants is worthwhile and should be encouraged* (Eurobarometer, 1992), national differences on issues like organ and tissue transplants, or assisted human reproduction were notable. For this reason, particular attention was paid to them when it came to articulating a European system of circulation of biological materials of human origin both for research purposes, and for biomedical application. Efforts were basically concentrated on defining

[39] EcoNexus, European Beekeping, CEO, Friends of the Earth, Gene Watch, Greenpeace, TestBiotech and European Coordination Via Campesina.

the common framework, and dealing with differences over the idea that between black and white there are many shades of grey. Respecting diversity, and assuming restrictions in use and regulatory nuances that are easily traced analysing comparatively the legislation of the 28 Member States, the free movement of human tissues and biobanks function satisfactorily in the EU, and form the basis today for the development of numerous biotechnological applications.

GMOs, however, have gradually been classified as black and/or white (Vázquez-Salat and Houdebine, 2013) and the public's perception of them has deteriorated, reflecting an increase in the number of people who see GM food *as risky, not useful and not morally acceptable* (see Eurobarometer surveys of 1991, 1993, 1996, 2002, 2005 and 2010 (Eurobarometer, 2010)). The evolution of this perception cannot, in my opinion, be separated from the controversy generated with in relation to GMOs, nor from the lack of effort on the part of Member States and Commission better to make public the advances in agrobiotechnology, and the benefits and risks associated with the latter (Lucht, 2015; Varzakas, 2007). The lack of trust in genetic engineering applied to other living organisms contrast with the significant number of people consulted in the surveys who, for example, are in favour of employing *genetic therapy* with humans, despite the fact that this technique, apart from being restricted is, in some cases, punishable as a crime.

12.4.2 It corresponds to Member States to address define and justify the particularities linked to GMOs on the basis of which, specifically, they exercise their opt-out

Although it is not said in exactly these words, the fact is that the third regulatory framework is presented in such a way that, the Commission attributes to Member States full responsibility when it comes to defining and justifying their opt-out. In some way, this allocation of responsibilities is also implicit in the way in which the EESC has expressed its position with regard to the proposal for a third regulatory framework for GMOs in the EU:

> Those Member States that have hitherto been determined that no decisions should be taken, or have been systematically opposed to sufficient majorities being formed, might no longer deem it necessary to mount political opposition within the Council if they can use other tools at national level to counteract or void the Council's decisions on this subject within their borders.

In this sense, says the EESC, the proposal *may remove an anomaly in the EU's legal practices* and at the same time, precisely because of that requirement that opt-out be justified and in accordance with certain limits, mechanisms have been provided for to avoid the misuse of this prerogative by Member States.

Some opinions have argued that the previous system did '*not allow the individual concerns of democratically elected governments to be taken into account*' (EPRS, 2015), such as for example, *questions not based on science* (EESC, 2015) or *other social considerations* (CoR, 2015), something that does seem to be included in the third regulatory framework. One thing, however, should be clarified. After the 2001 Directive, Member States, in the Standing Committee and the Appeal Committee, did have margin for expression of these concerns; what they did not have was: (1) the guarantee that the other Member States would accept that individual concern (the implicit consequence of any system of voting rules in a democracy); (2) or, where appropriate, an alternative guarantee of claiming for themselves a national exception. It is logical that in this respect, those that would have preferred a more extensive revision of the regulatory framework for GMOs express their discontent; thus, for instance, the CoR is of the opinion that *there has only been a revision of the use of GMOs* within national territories, as opposed to a radical revision of the authorisation procedure. But, as I have explained above, such a radical revision would not have been possible within the *core assumptions* around which the Commission has designed the third regulatory framework for GMOs. In the short term, and within the regulatory framework to be implemented, States will be allowed to decide with regard to certain particularities linked to GMOs. In the medium and long term, obviously, nothing prevents Member States that so wish from proposing new reforms of the regulatory framework and even, if appropriate, to the limits, which – established in the treaties – prevent the reform they seek to undertake.

The ad hoc justification of the restrictions on the use of GMOs.

The proposal for modification of Regulation 1829/2003 does not offer a list of possible compelling grounds to justify the restrictions on the use of GMOs, but Directive 412/2015 does, indicating that the measures must be in conformity with Union law, reasoned, proportional and non-discriminatory' and, in addition, 'based on compelling grounds such as those related to: (a) environmental policy objectives; (b) town and country planning; (c) land use; (d) socioeconomic impacts; (e) avoidance of GMO presence in other products without prejudice to Article 26a; (f) agricultural policy objectives; (g) public policy.

The grounds may be invoked individually or in combination, with the exception of the ground set out in point *(g) public policy* which cannot be used individually, depending on the particular circumstances of the Member State, region or area in which those measures will apply.

Those grounds, essentially, are a transcription of Article 36 Treaty on the Functioning of the European Union (TFEU) to Directive 412/2015, with the odd exception. Thus, an initial consideration to bear in mind is that article 36 TFEU itself, in its final clause, indicates that such prohibitions or restrictions '*shall not, however, constitute means of arbitrary discrimination or a disguised restriction on trade between Member States'* it should not be regarded therefore, as a novelty that, in order to exercise the opt-out in relation to GMOs, these requirements are also included: in conformity with Union law, reasoned, proportional and non-discriminatory. Meanwhile, and turning to the list of compelling grounds, the list's wording basically coincides with article 36 TFEU, which establishes the possibility of introducing prohibitions or restrictions on imports, exports or goods in transit

> justified on grounds of: public morality, public policy or public security; the protection of health and life of humans, animals or plants; the protection of national treasures possessing artistic, historic or archaeological value; or the protection of industrial and commercial property.

One of the few differences is that the Directive – unlike art. 36 TFEU – leaves no opt-out margin to Member States based on the criterion of 'protection of health and life of humans, animals or plants'. In a resolution of January 13, 2015 (EP, 2015) during the drafting of Directive 412/2015, the

European Parliament requested that environmental grounds be among the reasons which Member States could use to justify their cultivation bans (EPRS, 2015), but this was not accepted by Council, which deemed it to belong to the risk assessment tasks of EFSA. Consequently, in the terms in which the Directive was approved, environmental questions that could be alleged will have to differ from those related to risks and '*shall in no case conflict with the environmental risk assessment carried out [at EU level] pursuant this Directive or to Regulation (EC) No 1829/2003*'.

With the exclusion of scientific-technical risks, the grounds that can be employed to justify measures restricting the use of GMOs focus, as a whole, on socio-economic and environmental considerations, although in this respect there are also some comments and clarifications to be made. On the one hand, it is comprehensible that the range of possible grounds are of this nature, because risk concerns apart, this was the line followed by most of the concerns expressed by the Member States that proposed restrictions or prohibitions on GMOs. But, on the other hand, a significant problem is posed. These socio-economic and environmental grounds which may lead to restricted use of GMOs, precisely because they are grounds that are not linked to risks, have a very limited margin of application according to EU Law and jurisprudence. The ECJ has been interpreting narrowly the list of restrictions in art. 36 TFEU, tending to accept almost exclusively those measures which are seeing as related to non-economic interests (*Case C-120/95 Decker, 1998 (ECJ, 1988); Case 72/83 Campus Oil, 1984 (ECJ, 1984)*), and it remains to be seen what stance it will take with regard to opt-*out* requests vis-à-vis GMOs.

As well as the grounds, the specific measures adopted on the basis of these are also open to review

The Commission gives no '*indication of how countries can implement such a decision with legal certainty at national level*' and, very critical, the CoR Estimates that it is difficult to predict '*to what extent the prohibition option could be used at all in practice given the multitude of conditions to be met*'. But these same conditions must fulfil, in general, all the restrictions on or obstacles to the single market that a member state seeks to impose upon any type of product in any sector, because in the Treaties, as a rule, Member States have committed to a single market and to the principle of mutual recognition. I agree, in any case, with the EESC in its criticism of '*the lack of an exhaustive definition of the reasons that might justify the adoption of exclusion clauses – the absence of a positive or a negative list – as well as the lack of provision of legal mechanisms to suspend national measures that could be considered unfair, not sufficiently justified, or discriminatory, makes legal uncertainty a real risk*'. It warns that '*only the Court of Justice of the European Union will be able to resolve these uncertainties, unnecessarily tying up Member States' administrative work in legal battles and potentially causing delays and higher costs*'. Added to over two decades of moratoria, this new uncertainty, in my opinion, is not good for Member States, or for those participating in agrobiotechnological research and development nor for those operating as stakeholders in the EU food and feed chain.

An additional challenge to that of justification is that of defining the specific measures to be proposed as 'restrictions on use'. In the case of Directive 412/2015 this is not complicated, these are restrictions on cultivation, but the spectrum of restrictions is much wider with regard to the amendment of Regulation 1829/2003. However reasonable the justification, it is possible for the restrictive measure chosen, itself, to be ruled disproportionate or discriminatory, or for the ECJ to consider that the same objective reflected by the justification may be achieved in a less restrictive manner (AS. 155/82 *Com.*

v. Belgium, 1983). The fact that the proposal for amendment of 1829/2003 was not approved has postponed, in a way, reflection on this point although I would like to include here some observations on this question. If we examine ECJ jurisprudence, prohibitions o restrictions on use have been considered fundamentally on the basis of grounds (territorial, local) of security. For example, the ECJ upheld a general prohibition in Sweden on the industrial use of trichloroethylene because of the health and life of humans' justification (*Kemikalieninspektionen v. Toolex Alpha AB*, Case C-473/98, 2000, ECR I-5681, 49) and there have been cases related to road safety regulations that affect the use of specific vehicles. By way of example: the decision to ban on certain motorway routes in Austria the use of trucks of of more than seven and a half tons, carrying certain goods (*Commission v. Austria, Case C-320 03 (2005) ECR I-9871, 1*); the case *Commission v. Italy* regarding the prohibition on the towing of trailers by mopeds (which does not ban the acquisition of these vehicles in Italy, but obviously affects their sales); *Case C-142/05 Åklagaren v. Mickelsson and Roos*, on the use of personal watercrafts, and *Case C-265/06, Commission v. Portugal*, with regard to the prohibition on the attachment of coloured foils to the windows of motor vehicles for the transport of persons or goods (Gormley). It happens, however, that, in the case of the opt-out for GMOs safety concerns are excluded by the provision, as, practically, are those regarding labelling too, so Member States will have to find other alternative means of restriction that might have a place in the current regulatory framework.

12.4.3 The responsibility associated with the exercise of the opt-out prerogative will correspond, also exclusively, to the Member States exercising the latter

The affirmation with which I opened this section has not been expressly included in the proposal for reform of the regulatory framework for GMOs undertaken by the Commission or in the different opinions (EESC, CoR, European Parliament, etc.) that I have used as references, but neither can it be said that it is something that any of them have overlooked. Allowing Member States an opt-out based on criteria that are not related to the risk management of GMOs (which is still harmonised at EU level) and assigning them the responsibility of having satisfactorily to define and justify the measures via which they wish to exercise this prerogative has an obvious consequence: it will in the first instance be the responsibility of each Member State to assume any consequences that, as a result of its opt-out, may arise with regard to international agreements, fundamentally those relating to free trade, entered into with third countries. However, neither can the EU claim to be adopting a neutral position. Since the 1990s, in the document The Development and Future of the CAP, the Commission declared that '*the Community must recognise the existence of international interdependence and accept its responsibilities as the leading world importer and second leading exporter*' (COM 91 100, 11), and in coherence the CAP operates within an environment which includes the *Agreement on Agriculture*, the *Agreement on the Application of Sanitary and Phytosanitary Measures* (SPS), the *Agreement on Technical Barriers to Trade* (TBT), and the *General Agreement on Tariffs and Trade* (GATT, 1994). These Agreements, it is important to bear in mind, not only outline the contours of the CAP and each Members Agricultural policy, but also impose limits on domestic regulation (McMahon, 2007). It is the duty of the WTO to monitor its fulfilment and it must act as mediator in any dispute arising between member countries (147 countries in total). In case of conflict between countries, WTO arbitration shall prevail.

At the beginning of the century, and in relation to its agrobiotechnological policies, three claims were filed against the EU before the WTO arbitration panel: the cases of the United States (DS 291/17), Canada (DS 292/17) and Argentina (DS 293/17) against the European Communities. The WTO case against the EC regulatory regime came at a highly charged political moment, just after the expansion from 10 to 25 members (2004). The Dispute Settlement Body adopted three panel reports on three grounds: (1) the application of a general *de facto* moratorium on approval of GM products from June 1999 to August 2003; (2) the existence of undue delays with respect to 23 product-specific applications (out of the 27 cases considered by the Panel); and (3) national safeguard measures introduced by 6 Member States, which were found not to be based on an appropriate risk assessment.

The WTO arbitration panel interpreted in a very different manner (WTO, 2006) the practices of the EU and of the Member State that, according to the claimants, affected Biotech Products. On the one hand, and with regard to the delays in the approval procedures at EU level (1st and 2nd grounds), the WTO considered that unjustified delay could not strictly speaking be considered a breach of the *SPS Agreement* (WTO, 2006). However, and regarding the use of safeguard measures in force in 6 Member States (Austria, France, Greece, Germany, Italy and Luxembourg) under the Directives in force – 3rd ground –, the panel did consider that these restrictions violated *SPS Agreement* as they had not been justified by scientific evidence (Bodiguel and Cardwell, 2010; WTO, 2006). The WTO does not totally rule out the application of precautionary approaches in trade in GM crops and food GM of vegetable origin. Nevertheless, and compared with what appears to be the approach adopted in the Cartagena Protocol, what the WTO agreements do demand is clarity and precision in the application of precautionary criteria. The objective of this is to avoid, with the appearance of precaution, the application of what is essentially protectionist practices (Punt and Wesseler, 2016). Thus, the appellate body on measures affecting agricultural products has developed a series of constraints within which one might consider *reasonable* a measure based on a precautionary approach.

In the light of the DBS settlement (WTO, 2006), the EU and the three complainants (US, Argentina and Canada) agreed to engage in technical discussions on biotech-related issues and established November 2007 as a *Reasonable Period of Time* (RTP) for the implementation of agreement they reached to resolve the situation. Upon expiry of this period, the decision was made to extend the RTP a little, until 2008, and on its expiration the complainants adopted differing stances. Argentina and Canada resolved their differences with the EU in 2009 and 2010[40], but the dispute with the USA, to which new elements have been added, has not been resolved yet (WTO, 2015).

The incorporation of opt-out into the regulatory framework for GMOs may add, in the future, new elements to claims against the EU before the WTO. Given that risks cannot be included in compelling grounds, it will probably be the *TBT Agreement* (and not that of the *SPS Agreement*) which will be employed to decide whether or not the opt-out measures adopted by a Member State are compatible with international agreements on agriculture. The *TBT agreement* considers that no country should be prevented from establishing the measures necessary to guarantee the quality of exports, protect human, animal or plant health or prevent disloyal practices. However, it does

[40] EU and Canada settle WTO case on Genetically Modified Organisms. IP/09/1142. Brussels, 2009.

establish as a limit that these practices must always satisfy criteria of necessity and proportionality. So far, at a July 2015 meeting of the World Trade Organization's SPS Committee[41], Argentina, Paraguay, USA, Uruguay, Brazil and Canada reportedly contested the proposed new EU legislation allowing individual EU Member States to ban products without a risk assessment. This meeting took place, it should be noted, before 19 Member States exercised their opt-out. It remains to be seen whether now the justifications to which they have resorted – which in some cases are very superficial – will or will not comply with the terms of necessity and proportionality referred to by the *TBT Agreement*.

12.5 Will it be third time lucky? The glass half empty or half full, depending on how you look at it

The impasse preventing the use GMOs in the agri-food sector and, to a certain extent, any prospect of agrobiotechnological innovation within the EU, is much more serious than might at first appear. At the beginning of this work I said that I wondered, rather innocently perhaps, whether the proposal for partial devolution of the competence to Member States, itself exceptional and atypical of the trajectory of the EU, might not be somewhat *Solomonic* and, at the same time, whether one might expect this regulatory framework to prove to be a case of third time lucky. In the brief history of Directive 412/2015, for the time being the only law of this regulatory framework that has been passed, there has only been one legal consequence of note. The 19 Member States which, in their territory as a whole or part of it, had already *de facto* established restrictions on GMO cultivation, now and in the wake of Directive 412/2015, have seen formal recognition of that decision. In other words, a *de facto* situation, outside legality, is now recognised as a situation protected within EU Law.

However, beyond this minimal legal impact, it cannot be denied that from the social and political perspective the short but intense processing of these legal reforms has had a notable effect. There has been expression of a multitude of concerns y opposed views that, though they are not new, never before had they been reflected in writing and with such clarity, generating a new level of awareness when it comes to interpreting the current dissent with regard to GMOs in the EU. I do not honestly believe that the third regulatory framework will be the definitive version. It is difficult, within the rationale established by the Commission, to review the amendment of Regulation 1829/2003 or for the amendment, (in its present form) to be approved. But, in a sense, it has proved successful in some questions in relation to which the regulatory framework created in 2000 was not effective. The crisis of the first regulatory framework, of the 1990s, was interpreted fundamentally – although not only – as a problem of disagreement over management of risks to health and the environment (Escajedo San-Epifanio, 2008). In coherence, the rationale of its reform, in other words, the key areas of the second regulatory framework pointed to a full harmonisation of risk management at EU level. The practice demonstrated, however, in the background of the conflict there was much more than a question of how to manage risks and the third regulatory framework has changed the approach via which to contemplate dissent. In the words of the EESC, what it has done is offer tools so Member States who so desire may at national level counteract or void EU decisions on GMOs within their borders, without need to continue to block the rest via no opinions. Whenever the

[41] 5-16 July 2015 meeting of the World Trade Organization's SPS Committee, which deals with food safety and animal and plant health.

exercise of this opt-out is conditional upon the measures employed being reasoned and compatible with the principles of non-discrimination and proportionality, it seems to represent, at least, some progress compared to the previous situation. With acceptance of the fact that Member States *are not in agreement*, formally at least, a step has been taken towards revealing *where the disagreement is* and living together with the differences.

The context in which the new regulatory framework has been formulated, that of a Europe with many ongoing crises – political, economic and social – probably did not allow for a more ambitious approach with regard to GMOs. *Agenda 21* of the United Nations, one of the most important references of environmental legislation in force in most countries in the world, is a programme that neither flatly rejects biotechnology nor considers it to be immoral or unnatural. Moreover, warning of the need to temper *expectations with realism*, the Agenda indicates that environmentally sound management of biotechnology will contribute to improved *availability of food, feed and renewable raw materials*, '*by using the combined resources of modern biotechnology and conventional plant/animal/micro-organism improvement, including the more diverse use of genetic material resources, both hybrid and or original*'. But it also indicates, and this is probably the issue about which EU Member States disagree most, that biotechnology management faces, in the agri-food sector, the challenge of balancing '*the needs of farmers, the socio-economic, cultural and environmental impacts of modifications and the need to promote sustainable social and economic development*'. Member States do not share the same vision regarding hardly any of the elements to be balanced, and neither do they appear to agree over how to manage the coexistence between new and traditional agricultural techniques. They also have different views with regard to the pressure exerted upon them by the single market and imports from third countries, and one can easily imagine that in relation to many agri-food products there are always occasions when they would like to address socio-economic considerations that go (far) beyond the limited intervention imposed upon Member States by the obligations of the free market.

Against this swell, as I have said, the attempt to resolve the exceptional situation in which the regulatory framework for GMOs found itself cannot be successful if nothing other than technology is taken into account. Examining the difficulties involved in defining a common policy on the agri-food use of GMOs we encounter, amongst other things, the competing views held by Member States vis-à-vis the CAP and, even, as the EESC acknowledges, the increasing questioning on the part of some of '*the competences conferred by the Treaties in this area*'. If there were a common will, as there has been when agreeing upon such complex frameworks from the ethical-social and legal point of view as those concerning the movement of human tissues or the establishment and functioning of biobanks with biological materials of human origin, it would probably be possible to find a formula of coexistence between the common and the diverse, in which it might be possible to combine different ways of understanding an contributing to the environmentally sound management of biotechnology. In fact, in the environmentally sound management of *other technologies* it is also possible to detect different opinions amongst the 28 Member States and follow the example of the formulae employed in order to reach agreement. But here the problem lies in the fact that there is some doubt as to whether there exists a real spirit of understanding. This does not prevent the Commission, in its role, from invoking the so-called *loyalty clause* or *principle of sincere cooperation*, but in order to be successful a real desire to collaborate will be indispensable. If this is not the case, the Commission has warned, it does not have a *plan B*.

Time will tell where this route leads us. While the situation is clarified, I do not wish to end without indicating something to which, both in the crisis of the regulatory framework in the late1990s and in el current one, the Commission has not paid attention. Its efforts, as occurred in the 2000s, have been aimed at drawing towards consensus the Member States that were wary of modern biotechnology and has barely paid attention to the situation of those that participate in scientific-technical development or in agrobiotechnological companies. Neither have the opinions of the CoR or European Parliament paid excessive attention to this; perhaps, to some extent, did it the EESC. It is as if it had been thought, for the second time, that only by guaranteeing compliance with the provisions of the law – in other words, observance of the law *de facto* – would the interests of the parties involved in scientific-technical development be adequately protected. But what is lacking, perhaps, is some mention of the fact that the mere passing of laws so far and since the late 1990s, has brought neither legal certainty nor the fulfilment of the right to good administration (art. 41 CFREU) for researchers or stakeholders. Before, long before this situation, the European Council in Stockholm in March 2001 warned of the possible consequences of this situation and invited the Commission, together with the Council to '*examine measures required to utilise the full potential of biotechnology and strengthen the European biotechnology sector's competitiveness*'. Just a year later, the Commission admitted in the document *Life Sciences and Technology – A Strategy for Europe*[42] that '*Europe's current performance in life sciences and biotechnology is not facilitating the achievement of that objective*' (EC, 2002). Europe, it added, does not have a single policy for biotechnology but '*a patchwork of specific regulation, overlaid by many sectoral and horizontal policies at international, Community, Member State and local levels*'. If Europe is to successfully manage biotechnology and reap the benefits for society, said the Commission, we '*should proceed on the basis of a shared vision for a cooperative approach and with effective implementing mechanisms to compensate for absence of overall responsibility and control*', and we '*need to ensure coherence across Community legislation and policies directly regulating, or indirectly impacting on, the development of biotechnology*'. Because, '*without such mechanisms, biotechnology risks continuing to suffer indecision or short-sighted and local solutions*' (EC, 2002). A decade and a half later, with the situation still unresolved, biotechnology continues to suffer on account of this indecision. Let us hope that in the elements still missing from this third regulatory framework for GMOs or, who knows, perhaps in the foundations upon which the next one is based, attention is finally paid to those challenges to which – as well as to the risks – the agrobiotechnological policy of the EU should also provide a response.

Acknowledgements

Principal Researcher of the Collaborative research platform ELSA-SCITECH (University of the Basque Country); member of Multilevel Constitutionalism and Integration Diversity (Political, Legal and Cultural), research Group of the Basque Academic System IT743-13 and IT133-16, University of the Basque Country, Bilbao, Spain.

[42] *Life Sciences and Biotechnology – A strategy for Europe*, EC (2002) 27.

References

Alemanno, A. (2013). Science, law and policy of neonicotinoids and bees: a new test case for the precautionary principle. European Journal of Risk Regulation 2.

Ammann, N. (2014). Genomic misconception: a fresh look at the biosafety of transgenic and conventional crops. A plea for a process agnostic regulation. New Biotechnology 13(1): 1-27.

Bernauer. T. (2003). Genes, trade and regulation. Princeton University Press, Princeton, NJ, USA.

Bodiguel, L. and Cardwell, M. (2010). The regulation of genetically modified organisms: comparative approaches. Oxford University Press, New York, NY, USA.

Bratsch, D. (2014) GMO regulatory challenges and science: a European perspective. Journal für Verbraucherschutz und Lebensmittelsicherheit 9:S1.

Byk, C. (2008). The nexus of law and biology. new ethical challenges. Ashgate Publishing, Farnham, UK.

Committee of the Regions (CoR) (2015). Opinion on the decision-making process on genetically modified food and feed, NAT-VI/003. Opinion No. CDR 3636/2015: 12-14.

Cook, J. and Lewandowsky, S. (2011). The debunking handbook. University of Queensland, St. Lucia, Australia.

De Francesco, L. (2013). How safe does transgenic food need to be? Nature Biotechnology 31, 794-802 (2013).

ECJ (1984). Case 72/83 – Campus Oil Limited and others v. Minister for Industry and Energy and others. Available at: http://tinyurl.com/yapk79ae.

ECJ (1998). Case C-129/95 – Decker v. Caisse de maladie des employés privés. ECLI:EU:C:1998:167.

Escajedo San-Epifanio, L. (2012). Transformation of food governance models: perspectives arisen from a food citizenship. In: Potthast, T. and Meisch, S. (eds.) Climate change and sustainable development. Ethical perspectives on land use. Wageningen Academic Publishers, Wageningen, the Netherlands.

Escajedo San-Epifanio, L. (2008). Biotechnologie, santé et environnement dans l'Union européenne: aspects politiques et juridiques. Revue du Marché commun et de l'Union Européenne, No. 517.

Escajedo San-Epifanio, L. (2010). Vers un cadre regulateur de la sécurité de l'application de la biotechnologie: le chemin parcouru. Revue Générale de Droit International Public 114: 5-34.

Escajedo San-Epifanio, L. (2015). Challenging food governance models: analyzing the food citizen and the emerging food constitutionalism from an EU perspective. Journal of Agricultural and Environmental Ethics 26(3): 435-454.

Eser, A. (2002). Auf der Suche nach dem mittleren Weg: Zwischen Fundamentalismus und Beliebigkeit? In: Langer, M. and Laschet, A. (eds.) Unterwegs mit Visionen. Festschrift für R-Süssmuth, Freiburg, Germany, pp. 139.

Eurobarometer (1992). Public opinion in the European community, n° 38, December 1992. Available at: http://tinyurl.com/yablx768.

Eurobarometer (2010). Eurobarometer surveys of 1991, 1993, 1996, 2002, 2005 and 2010. Available at: http://tinyurl.com/y9p9vh4x.

European Commission (EC) (2000). Communication from the Commission on the precautionary principle. COM (2000) 1 final, February. EC, Brussels, Belgium.

European Commission (EC) (2002). Communication from the Commission to the European Parliament, the Council, the Economic and Social Committee and the Committee of the regions, life sciences and biotechnology – a strategy for Europe, COM (2002) 27. EC, Brussels, Belgium.

European Commission (EC) (2015a). Communication from the Commission to the European Parliament, the Council, the European Economic and Social Committee and the Committee of the Regions. Reviewing the decision-making process on genetically modified organisms (GMOs). COM (2015) 176 final. EC, Brussels, Belgium.

European Commission (EC) (2015b). Proposal for a regulation of the European Parliament and of the Council amending Regulation (EC) No. 1829/2003 as regards the possibility for the member states to restrict or prohibit the use of genetically modified food and feed on their territory. COM/2015/0177 final – 2015/0093 (COD). EC, Brussels, Belgium.

European Economic and Social Committee (EESC) (2015). NAT/669 – EESC-2015-02913-00-00-AC-TRA (EN) 177-178. EESC, Brussels, Belgium.

European Parliament Research Service (EPRS) (2015). Imports of GM food and feed. Right of member states to opt out. Briefing – EU legislation in progress – 19 October 2015, PE 569.030. EPRS, Brussels, Belgium.

European Parliament (EP) (2015). European Parliament legislative resolution of 13 January 2015 on the Council position at first reading with a view to the adoption of a directive of the European Parliament and of the Council amending Directive 2001/18/EC as regards the possibility for the member states to restrict or prohibit the cultivation of genetically modified organisms (GMOs) in their territory (10972/3/2014 – C8-0145/2014 – 2010/0208(COD)). EP, Brussels, Belgium.

Faulkner, A., Lange, B. and Lawless, C. (2012). Material worlds: intersections of law, science, technology and society. Journal of Law and Sociology 39(1): 1-19.

Gormley, L.W. (2001). Free movement of goods and their use – what is the use of it? Fordham International Law Journal 33(6): 1587-1628.

International Service for the Acquisition of Agri-biotech Applications (ISAAA) (2015). Global status of commercialized biotech/GM crops: 2015. ISAAA Brief No. 51. ISAAA, Ithaca, NY, USA.

Klümper, W. and Qaim, M. (2014). Meta-analysis of the impacts of genetically modified crops. PLoS ONE 9(11): e111629.

Levy, D. and Newell, P. (2000). Oceans apart? Business responses to global environmental issues in Europe and the United States. Environment 42(9): 8-20.

Lucht, J.M. (2015). Public acceptance of plant biotechnology and GM crops. Viruses 7: 4252-4281.

McMahon, J. (2007). EU agricultural law. Oxford University Press, New York, NY, USA.

Müller-Röber, B., Boysen, M., Marx-Stölting, L. and Osterheider, A. (2013). Agricultural biotechnology. Current scientific, economic and societal developments. Supplement of the interdisciplinary research group 'Gene Technology Report'. Berlin-Brandenburg Academy of Sciences and Humanities, Berlin, Germany.

NGO coalition (2015). Open letter to the Commission on new genetic engineering methods. Available at: http://tinyurl.com/jwrb28c.

O'Riordan, T., Cameron, J. and Jordan, A., (2001). The evolution of the precautionary principle. In: O'Riordan, T., Cameron, J. and Jordan, A. (eds.) Reinterpreting the precautionary principle. Cameron May, London, UK.

Patterson, L.A. (2000). Biotechnology policy. In: Wallace, H., Pollack, M.A. and Young, A.R. (eds.) Policy Making in the EU, 4th edition. Oxford University Press, New York, NY, USA.

Poli, S. (2015). The reform of the EU legislation on GMOs: a journey to an unknown destination? European Journal of Risk Regulation 6: 559.

Pollock, C.J. and Hails, R.S. (2014). The case for reforming the EU regulatory system for GMOs. Trends in Biotechnology 32: 63.

Punt, M.J. and Wesseler, J. (2016) Legal but costly: an analysis of the EU GM regulation in the light of the WTO trade dispute between the EU and the USA. World Economics 39: 158-169.

Raybould, A. and Mitchell, P. (2003). Europe angers US with strict GM labelling. Nature Biotechnology 21(1): 6.

Rosendal, G.K. (2005). Governing GMOs in the EU. Global Environmental Politics 5(1): 82-92.

Salvi, L. (2016). The EU regulatory framework on GMOs and the shift of powers towards member states: an easy way out of the regulatory impasse? European Food and Feed Law Review 11(3): 201-210.

Shiemann, J. and Hartung, F. (2015). EU perspectives on new plant-breeding techniques. In: Eaglesham, H. and Hardy, R.W.F. (eds.) NABC report 26. New DNA-editing approaches: methods, applications and policy for agriculture. Boyce Thompson Institute, Ithaca, NY, New York, USA.

Spaak Committee (1956). Brussels report on the general common market. Available at: http://tinyurl.com/j6dx9eq.

Sprink, T., Eriksson, D., Shieman, J. and Hartung, F. (2016). Regulatory hurdles for genome editing: process- vs product-based approaches in different regulatory contexts. Plant Cell Reports 35: 1493-1506.

Sunstein, C.R. (2005). The laws of fear: beyond the precautionary principle. Cambridge University Press, Cambridge, UK.

UNESCO (2005). Report of UNESCO's group of experts on the precautionary principle. UNESCO, Paris, France.

United Nations Conference on Environment and Development (UNCED) (1992). Rio de Janeiro, 3-14 June 1992. Available at: http://tinyurl.com/yb2w4mj5.

Van der Meer, P. (1999). Environment sessions report. International Conference on Ethical Issues arising from Biotechnologies, Proceedings I. Available at: http://tinyurl.com/gl7x3pf.

Váquez-Salat, N. and Houdebine, L.M. (2013). Society and GMOs – chicken and egg? The lessons from the GM crops debate in Europe over risk management and communication provide valuable pointers for the upcoming debate on GM animals. EMBO Reports 14: 671-674.

Varzakas, T.H., Arvanitoyannis, I.S. and Baltas, H. (2007). The politics and science behind GMO Acceptance. Critical Reviews in Food Science and Nutrition 47: 335-361.

Von Homeyer, I. (2002). Institutional change and governance in the European Union: the case of biotechnology regulation. European University Institute, Florence, Italy.

World Trade Organization (WTO) (2006). European communities – measures affecting the approval and marketing of biotech products. Reports of the Panel. World Trade Organization, Geneva, Switzerland.

World Trade Organization (WTO) (2015). Dispute DS291 European communities – measures affecting the approval and marketing of biotech products. Available at: http://tinyurl.com/zc7hla8.